Fossil Plants from Wales

Michael G. Bassett
Department of Geology, National Museum of Wales, Cardiff

and

Dianne Edwards
Department of Plant Science, University College, Cardiff

AMGUEDDFA GENEDLAETHOL CYMRU
NATIONAL MUSEUM OF WALES

Geological Series No. 2

CARDIFF, OCTOBER 1982

ISBN 0 7200 0265 6

Designed by Penknife
Typeset by Afal
Printed by South Western Printers

Cover illustration
Gosslingia breconensis, a fossil plant from the Lower Old Red Sandstone (Senni Beds, Lower Devonian) of the Brecon Beacons, approximately x2 (by permission of the Trustees of the British Museum (Natural History)).

Note
This booklet is a revised and expanded version of an article that was first published under the same title in *Amgueddfa, Bulletin of the National Museum of Wales* No.13, Spring 1973, pp.2 – 27.

Palaeobotany, the study of fossil plants, provides a fragmentary and tenuous evolutionary picture of plant life from as far back as about 3,500 million years ago; that it presents any kind of picture is remarkable in view of the delicate nature of most plants, which would normally be expected to break up and decompose rapidly before or during the process of fossilisation, and it is clear that the plant fossil record must be even less complete than that of most groups of animals. It is not surprising, therefore, that in Wales, as in many other parts of the northern hemisphere, past work in palaeobotany has been dominated by studies of the floras of the late Carboniferous Coal Measures, where unique conditions supported a rich plant life and allowed much of it to be fossilised in fine detail; the economic importance of the coal-bearing strata, and the consequent extraction of vast amounts of rock which contained fossil specimens, added further stimulus to these studies.

However, over the last 20 years or so, increased attention has been directed throughout the world towards pre-Coal Measures plants, focussing in particular on two separate but equally important steps in botanical history within different parts of the geological column. The first of these steps involved the origin of life itself, with current discussion centred on the interpretation of the significance of microscopic plant assemblages (including bacteria and simple algae) that are now known from Precambrian rocks well over 3,000 million years old. The second step involved the origin and diversification of vascular land plants during Silurian and early Devonian times, approximately 435 to 390 million years ago. In Wales there are no rocks provably older than about 600-700 million years, so that the area provides no direct evidence relating to the origin of life, but in contrast, rocks from both Wales and the adjacent Welsh Borderland yield specimens which are proving to be particularly important in tracing the origins of vascular land plants, as well as in recording a number of subsequent steps in plant evolution.

With the increasing emphasis on the investigation of fossil plants in Wales, palaeontologists who previously recorded material as 'unidentifiable carbonaceous fragments' or 'plant debris' are becoming aware of the potential importance of such specimens and are passing them on to specialists for examination, while the reinterpretation of known species which were inadequately studied in the past is also proving to be rewarding. The result is that a great deal of new information has been obtained in recent years, and it seems certain that continued searching will reveal further new and exciting discoveries.

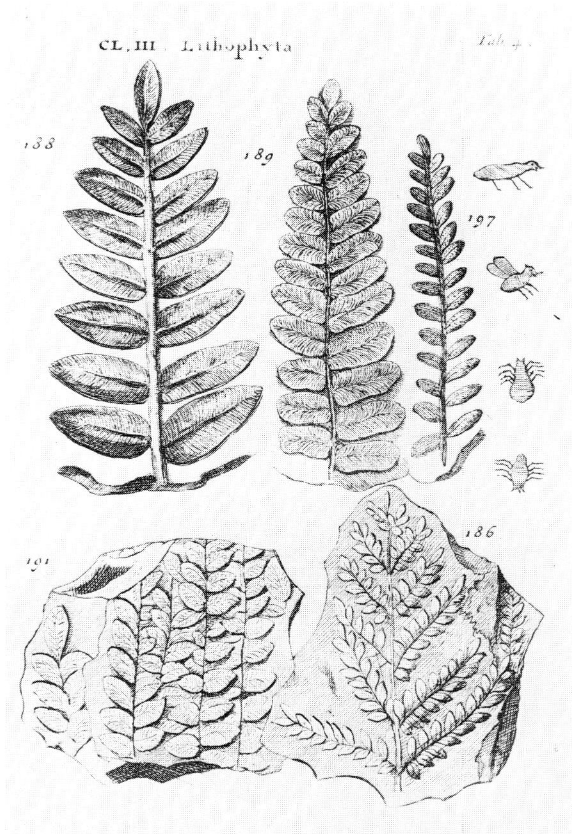

One of Edward Lhwyd's illustrations of fossil plants from the Coal Measures of Wales

It is also appropriate to focus attention on fossil plants from Wales for historical reasons, since specimens from the Coal Measures of North and South Wales were among the earliest to be described and illustrated; these were published in 1699 by the Welsh naturalist Edward Lhwyd, Keeper of the Ashmolean Museum at Oxford, who compiled the first ever catalogue of fossils entitled *Lithophyllacii Britannici Ichnographia.* Although Lhwyd did not appreciate the true nature of fossils as once living organisms, his illustrations were so good that in some cases the genera and species can be identified from them.

Early algae

A distinctive feature of present landscapes over much of the earth is a general abundance of vegetation with its predominance of green colours, but before about 400 million years ago the land surface was either strikingly bare or perhaps covered partially with a thin green algal scum. Evidence from the oldest fossils yet discovered, from very early

Precambrian rocks in areas such as Western Australia, North America, and South Africa, suggests that plant life began in the sea or possibly in coastal lagoons or on tidal mud flats, and under fairly hot climatic conditions. From these origins until about mid Silurian times the only convincing plant records are of aquatic forms.

The algae are a widely diverse group of predominantly aquatic plants, ranging from minute planktonic forms such as diatoms to the familiar seaweeds of our sea shores; they have a comparatively simple internal structure with the majority composed of soft tissue which is even more delicate than in higher terrestrial plants, but in a few algal groups the cellular framework is impregnated with silica or calcium carbonate ($CaCO_3$) and is thus more likely to become fossilised. Calcareous algae are known to play an important part in rock building, both in secreting calcium carbonate and in binding and trapping limestone sediments, and these kinds of biological/sedimentary interactions can be traced back into the early Precambrian in structures known as stromatolites.

Modern stromatolites are built by communities of blue-green algae (also called cyanophytes or cyanobacteria) and bacteria in warm, intertidal and very shallow subtidal environments (maximum recorded depth about 3.5 metres). Externally they appear as conical, columnar, or hemispherical mounds of sediment covered by an irregular coating of algae. In longitudinal section they are coarsely to finely laminated, with dark layers of organic-rich material (the remains of the communities) alternating with lighter layers in which sediment predominates. During sedimentation, surface layers are covered by mats of micro-organisms, usually filamentous blue-green algae, which trap the sediment and precipitate $CaCO_3$ that acts partly as a cement. The algae are capable of limited gliding movement, and shed their protective sheaths to move upward between the sediment particles towards the light, where they form a new surface layer. The dark layers therefore represent the abandoned sheaths, and as the process continues the laminated deposits build up into various shapes. Secondary alteration in fossil forms often destroys the organic structure of the algal layers, so that no direct biological evidence remains and the mounds are therefore not fossils in the strictest sense, but from the gross morphology and remnant internal lamination it is often possible to deduce a sedimentary/algal stromatolitic origin.

Stromatolites occur throughout the geological column but were apparently much more common in Precambrian times, probably because there were then no grazing gastropods to eat the algae. Limestones in the late Precambrian or early Cambrian Gwna Group to the east of Cemaes Bay and also south of Carmel Head, northern Anglesey, contain poorly preserved stromatolites that represent the oldest forms known in Wales. The assemblage includes morphological types similar to those named elsewhere as *Conophyton* and *Georginia*; the mounds reach up to 60 cm in diameter, with individual columns up to 7 cm wide. Slightly older beds of shale and jasper in the Gwna Group at Llanddwyn have yielded poorly preserved, filamentous and spheroidal organic-walled microfossils, but they are rare, difficult to interpret, and have not yet been studied in detail. However, from initial studies it has been suggested that the filamentous forms might represent empty sheaths of blue-green algae or, more probably, iron bacteria, while the spheroidal bodies resemble a problematical group known as actritarchs, some of which are possibly cysts of simple algae. Should these relationships be confirmed by further investigation, then the Llanddwyn microfossils will represent the earliest Welsh floras.

Rock building calcareous algae became

Stromatolites in outcrop of the Gwna Group, south of Porth yr Hwch, Carmel Head, Anglesey; note the distinctive lamination, becoming increasingly domed upwards (x 0·3 approx.)

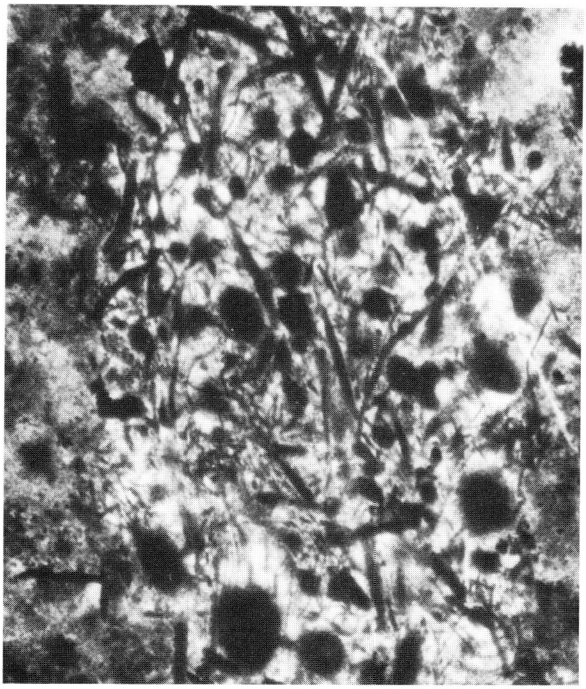

Thin section of micro-organisms in jasper from the Gwna Group, Anglesey; the filaments and spheroids are preserved as finely granular iron minerals (x 240)

increasingly diverse through Lower Palaeozoic times, although in Wales, where organic limestones are virtually unknown from the Cambrian period, they are next recorded as fossils in upper Ordovician rocks. Two major groups of the green algae, the Dasycladaceae and Codiaceae, together with the Solenoporaceae of the red algae and a few genera of less certain affinities but probably related to the greens or blue-greens, constitute the main calcareous forms. Of these the oldest reported Welsh genus is the dasycladacean *Coelosphaeridium* from Caradoc limestones of the Bala district, while in approximately contemporaneous rocks at Hoar Edge in Shropshire the red *Solenopora* and the long ranging green or blue-green genus *Girvanella* are present. During Silurian times all the groups became more prolific, notably in the well known Woolhope, Much Wenlock, and Aymestry limestones, but also in more restricted calcareous developments at other horizons. In the Llandovery Series the green algae are represented by *Rhabdoporella* and *Ortonella*, which are found together in limestone lenses in basal sandstones of Clwyd, and by the distinctive *Cyclocrinites favus* which first occurs near the base of the Series in the Gasworks Mudstone of the Haverfordwest district. *Solenopora* becomes especially dominant in the lower Wenlock, where it forms much of the massive algal limestone development of Dolyhir and Nash Scar in Powys in association with *Girvanella, Wetheredella* and *Rothpletzella.* The same association occurs in the Woolhope Limestone Formation, and all three genera play an important part in the construction of the reef or 'ballstone' structures of the Much Wenlock Limestone Formation. The upper surfaces of many fossils in the reefs are encrusted by stromatolites, while in the surrounding bedded limestones some of the algae form spheroidal structures known as oncolites.

Although not identical to living genera and species the general construction of the main fossil calcareous algae allows them to be related to modern forms with a good degree of confidence. One distinctive exception is the problematical *Ischadites koenigii,* a pear-shaped or globose species found in the Wenlock and Ludlow rocks of the Welsh area, first described by Sir R.I. Murchison in 1839 in *The Silurian System* as being of unknown affinities but possibly a mollusc, and subsequently assigned to a number of organic groups including crinoids,

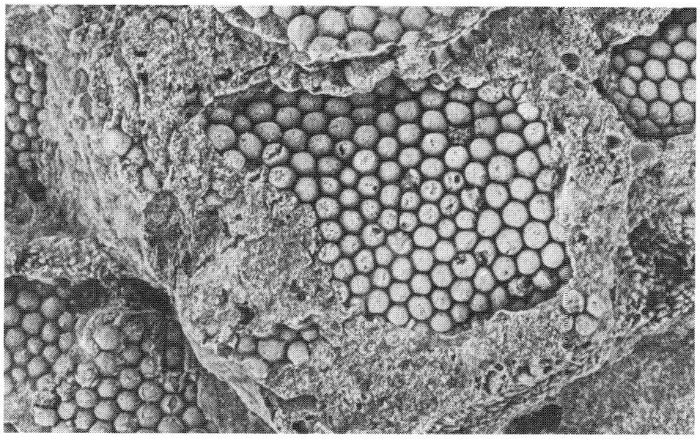

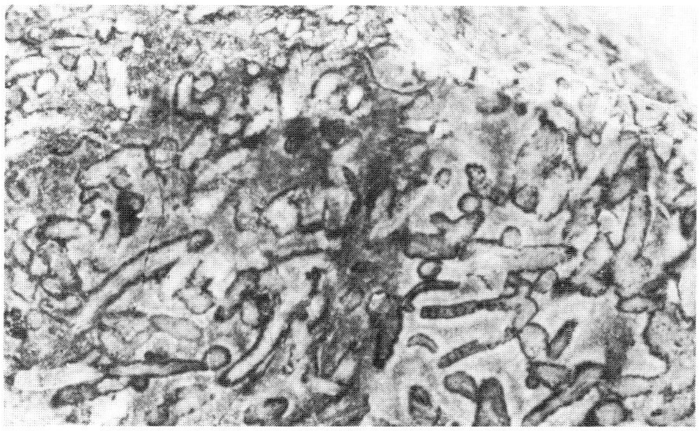

Silurian calcareous algae; **(above)** *Mould of colonies of* Cyclocrinites [Mastopora], *Lower Llandovery, Gasworks Mudstone, Dyfed.* **(below)** *Polished section showing algal tubes, mainly* Solenopora gracilis *and* Rothpletzella *sp., lower Wenlock, Dolyhir and Nash Scar Limestone, Dolyhir, Powys (x 1.5 approx.)*

Ischadites koenigii, *the original illustration of the species in* The Silurian System *(x 0.7 approx.)*

cystoids, sponges and dasycladacean algae; most authors now accept an algal interpretation, although a relatively recent account of related Ordovician forms from North America again proposed closer affinities with the sponges, with a suggestion that they may even represent a separate group with sponge and algal characteristics.

Among the non-calcareous algae relatively few are known as fossils, and those that are remain difficult to interpret since they are preserved as carbonaceous compressions. In Wales the earliest forms, referred to the genera *Eophyton* and *Buthotrephis*, were described from south-west Dyfed in the 19th century from rocks of late Cambrian and early Ordovician age (approximately 520-490 million years old), but it is possible that some of the included structures represent burrows or tracks of animals. Undoubted plants occur in the Silurian, from where two particularly interesting new discoveries were made in the early 1970s. The first, described in 1971 as a new genus named *Inopinatella,* was collected from the Aymestry Limestone of Shropshire; it is a simple, branching form of uncertain affinities, but it has been suggested that despite being non-calcareous it may represent a primitive type of dasycladacean, possibly related to the living tropical genus *Neomeris* which also has an initial uncalcified growth stage. The second discovery comprised exceptionally well preserved specimens collected in 1972 from lower Ludlow rocks at Llangammarch Wells, Powys, and described in 1977 as a new genus named *Powysia.* The most complete specimen consists of a thallus differentiated into a basal holdfast, stipe and much branched distal region, all of which appear to have a tubular construction of intertwining filaments or siphons. Precise affinities are again uncertain, but the gross morphology of *Powysia* suggests a benthic alga, probably of marine origin, although the possibility that it was a fresh water plant which was swept out to sea before burial cannot be ruled out.

Powysia bassettii, *a non-calcareous branching alga from the Silurian (Ludlow) of Llangammarch Wells, Powys (x 1.4 approx.)*

One of the most problematical non-calcareous plants is *Prototaxites*, which in Wales, as in other parts of the world, ranges through the Silurian and Devonian, the first Welsh examples being described from the Wenlock of Pen y Glog in Clwyd and Rumney near Cardiff. Its axes, over half a metre across in some Canadian specimens but more commonly confined to a few millimetres, consist of cylindrical tubes of two sizes and the generally accepted interpretation is that it is an alga, possibly related to brown algae such as the giant sea kelps. Some authors have suggested that *Prototaxites* is of terrestrial rather than marine origin and that this was an 'experiment in evolution', successful in the colonisation of the land in the Silurian and Lower Devonian but failing later and becoming extinct during Upper Devonian times; part of the evidence for such suggestions is based on the association of *Prototaxites* with undoubted vascular land plants in

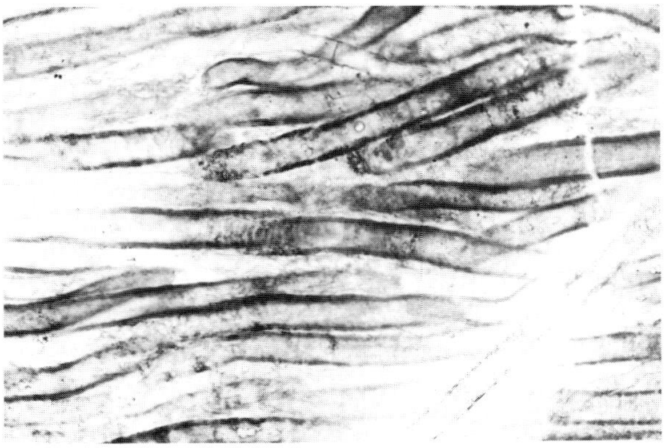

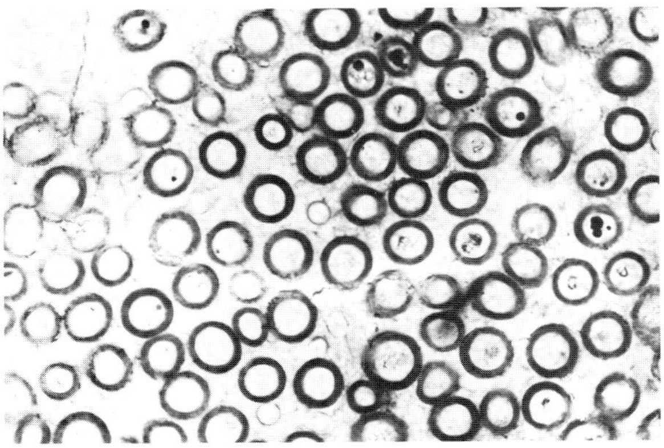

Longitudinal (above) *and transverse* (below) *sections through a petrified cylindrical axis of* Prototaxites storriei. *Only the larger tubes are visible. Silurian (Wenlock Series), Rumney, Cardiff.* (*x 150 approx.*)

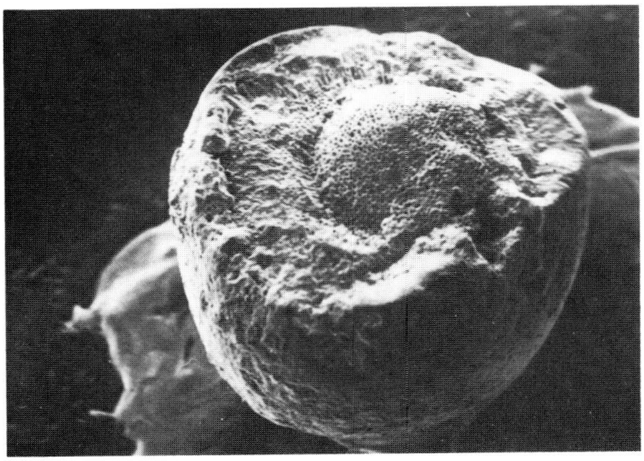

Scanning electron micrograph of Pachytheca *sp., broken open to show the two main internal layers; the central area comprises a mass of intertwining tubes and is surrounded by a further system of tubes which radiate outwards. Ditton Series, Welsh Borderland (x 30 approx.)*

the Downton rocks of the Welsh Borderland, but the interpretation is still regarded as speculative since separate floras could have been washed into the same area after death. A further problematical organism occurring with *Prototaxites* both in the Wenlock of North and South Wales and in Downton strata, as well as in isolation elsewhere in the Silurian, is *Pachytheca,* a small spherical body comprising a central area of interwoven tubes with a further system of tubes radiating out to the margins; it compares with some blue-green algae and may have rolled about on the bottom in a fresh or brackish water environment, although recent geochemical and morphological studies on specimens from the Lower Devonian of Scotland imply relationships with green algae (Chaetophoraceae).

Of all these early algal and probable algal groups, the green algae appear to be the most likely ancestors of the vascular land plants since extant representatives of both have the same types of pigments, mainly in the form of chlorophyll, both store food as starch, and both contain cellulose as a major component of the cell walls; there is, however, still no direct fossil evidence for the link between the two groups.

Invasion of the land

The first northern hemisphere record of an undoubted vascular land plant, named *Cooksonia,* is from the Downton Series (late Silurian) of Perton near Woolhope in Herefordshire. This was described by W.H. Lang in 1937 as a small, erect plant with naked, simple dichotomously branching axes and globular, terminal sporangia.

A vascular plant is one which possesses a water-conducting system composed of tracheids or vessels (tubular, dead cells) which transfer water from the absorbing regions to the rest of the plant. The majority of terrestrial plants such as ferns, conifers and flowering types are vascular. Such a conducting system is not a characteristic of the algae, which are normally surrounded by water; in fact the vascular system is considered to have evolved following the invasion of the land. Further morphological features associated with the land habit (i.e. modifications for existence in an environment in which water is not always readily available) are a waterproof, waxy covering or cuticle, pierced with small holes (stomata) for gaseous exchange, and spores with a tough, protective coating and typical 'Y' shaped (trilete) markings reflecting production in a tetrad, held high enough in small sacs (sporangia) to allow liberation into air currents; these additional modifications are not individually unique to vascular plants, but when found together can generally be regarded as diagnostic. In addition to tracheids,

Cooksonia cambrensis *from the Downton Series of Freshwater East, Dyfed. Note the simple forking axis ending in globose sporangia. The pin on the left gives an indication of the extremely small size (x 10 approx).*

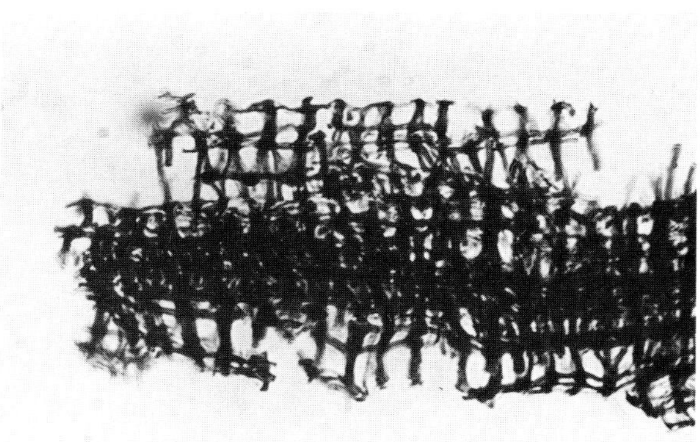

Isolated tracheids from the central water-conducting strand of a Lower Devonian plant from the Brecon Beacons. The horizontal and vertical bars were reinforced by lignin (x 150 approx.)

Cooksonia displays most of these characters and is clearly a vascular land plant.

The presence of such vascular plants (with their modifications for survival in terrestrial environments) in upper Silurian sediments is thus excellent evidence that by that time floras were established on the land as well as in water. Considerable effort has been put into the search for early land plants, much of it centred on Silurian rocks in South Wales, where it has involved the detailed investigation of the most fragmentary fossils for any signs of tracheids, spores and cuticles. Some of the earliest axes containing tracheids come from uppermost Ludlow strata at Capel Horeb in south-west Powys. Smooth, sterile, forkings axes are called *Hostinella*. At Capel Horeb these occur with isolated sporangia of *Cooksonia* and *Steganotheca*; the latter closely resembles *Cooksonia* in having simply branching smooth axes and terminal sporangia, but in detail the sporangia differ in being more elongated with parallel sides, truncated apices, and bases tapering smoothly down into the axes. The organisation in both these genera is as simple as it is possible to imagine in vascular plants. *Cooksonia* itself is also known from even older Ludlow strata at Cwm Graig Ddu in Powys, but as yet it has been impossible to detect vascular tissues or indeed any traces of anatomical structures in these compression fossils; similar material has also been described recently from Wenlock strata in Ireland, pushing the possible origins of vascular plants back into at least middle Silurian times.

Plants with the *Cooksonia* type of organisation have been found in Downton rocks throughout the

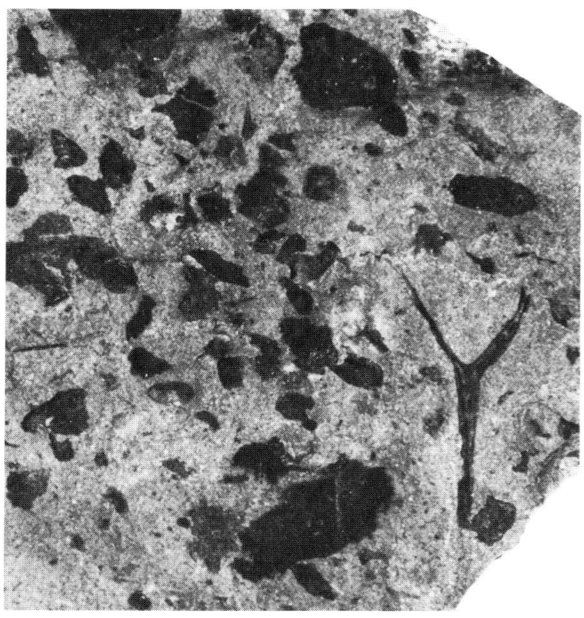

Typical rock surface from upper Ludlow (Whitcliffian) strata at Capel Horeb, Powys showing a branching axis (Hostinella) *and patches of black coalifield material which may have originated either from animals such as eurypterids or from encrusting plants (x 2)*

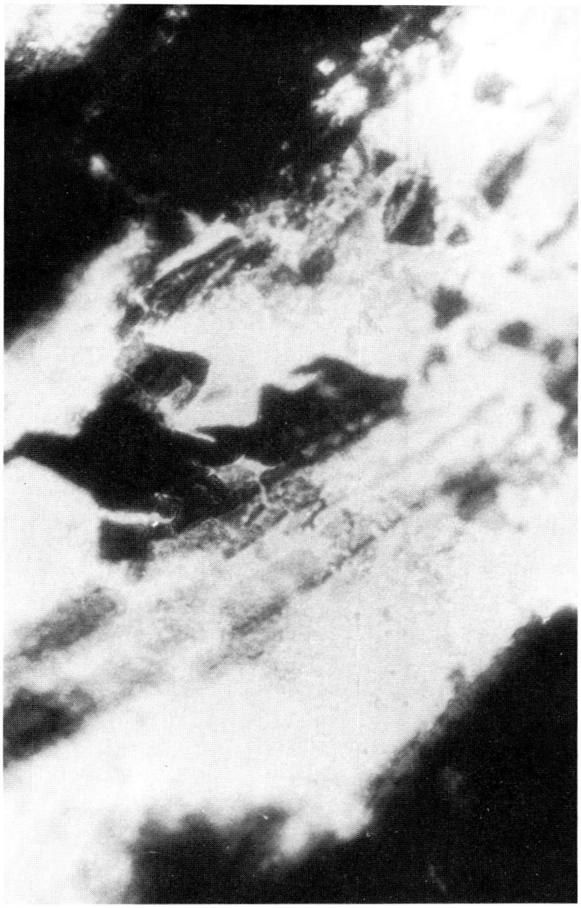

The earliest tracheids in Wales isolated from an unbranched axis in Whitcliffian rocks at Capel Horeb, Powys. The affinities of such a fragmentary fossil plant are unknown, but may have been with Cooksonia *or* Steganotheca *(x 490 approx.)*

world, in Podolia, Czechoslovakia, Germany, North Africa and North America. An intensively studied assemblage from Freshwater East, Dyfed indicates that land vegetation was beginning to diversify towards the end of the Silurian. Here, in addition to a variety of forms of *Cooksonia*, there is a new fertile genus named *Tortilicaulis*; in this plant, a smooth, unbranched axis which often appears twisted, terminates in a single elongate sporangium. The genus thus resembles the sporing bodies of certain modern liverworts, but the complete absence of preserved anatomy makes it impossible to distinguish between bryophyte and tracheophyte levels of organisation in this case. Not all of the axes in the assemblage are smooth; a bifurcating example bears small, triangular spine-like outgrowths.

By early Ditton times (395-390 million years ago) a second type of land plant architecture had evolved

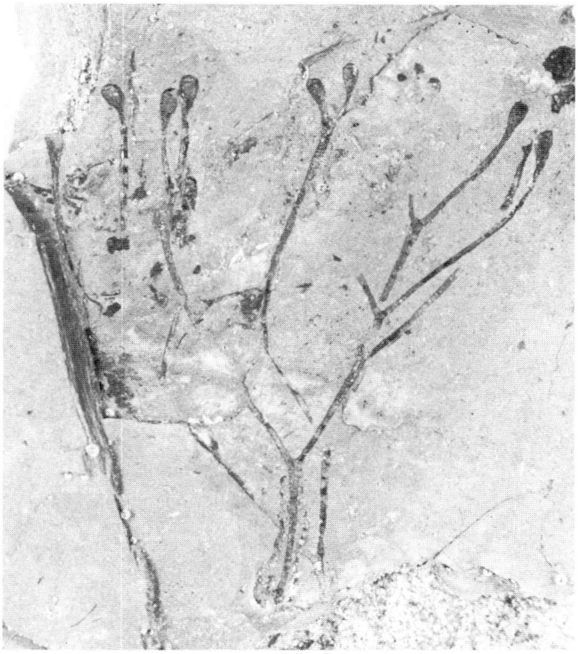

Steganotheca striata, *Downton Series, Capel Horeb, Powys. Note the slightly elongate sporangia with flattened tips. More fragmentary specimens have been collected together with* Cooksonia *from older Ludlow rocks at the quarry (x 1.5 approx.)*

Cooksonia *sp., Silurian (Ludlow Series, Bringewoodian), Cwm Graig Ddu, Powys. Note the smooth, forking axis with compressed terminal sporangia on the two left hand branches(x 4)*

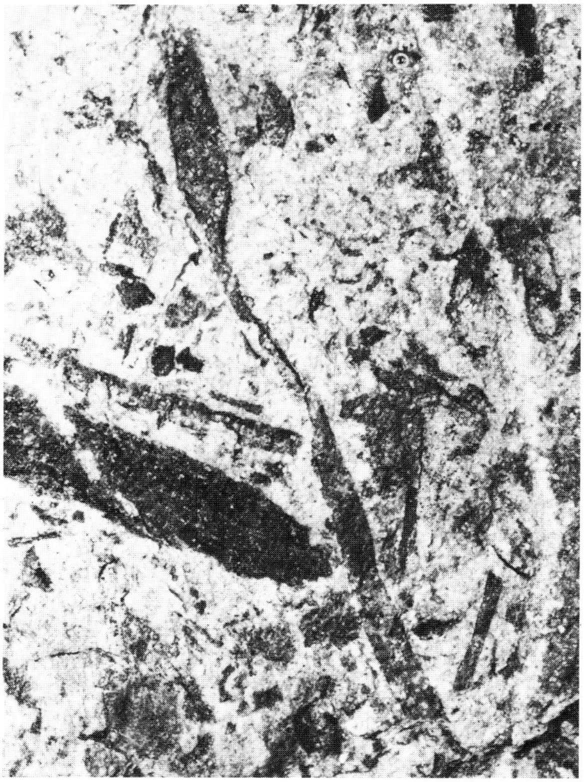

Tortilicaulis transwallensis. *Elongate, cigar shaped sporangium terminating a smooth, twisting axis. Downton Series, Freshwater East, Dyfed (x 10 approx.)*

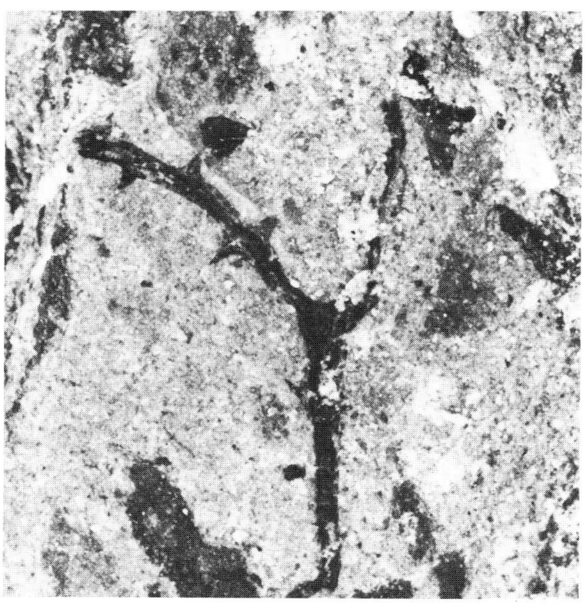

Forking plant axis; the left hand branch has conspicuous triangular outgrowths. Downton Series, Freshwater East, Dyfed (x 10 approx.)

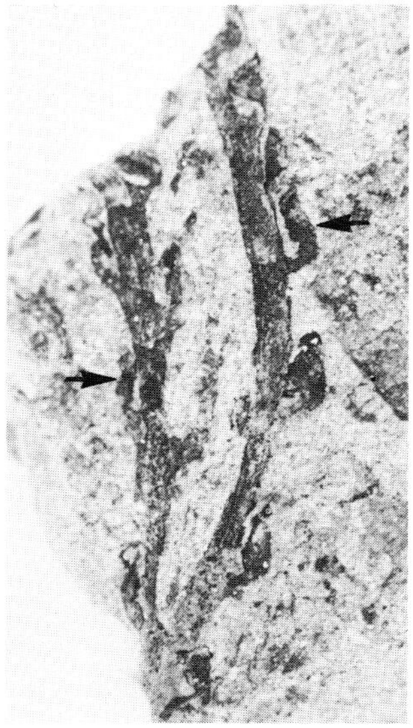

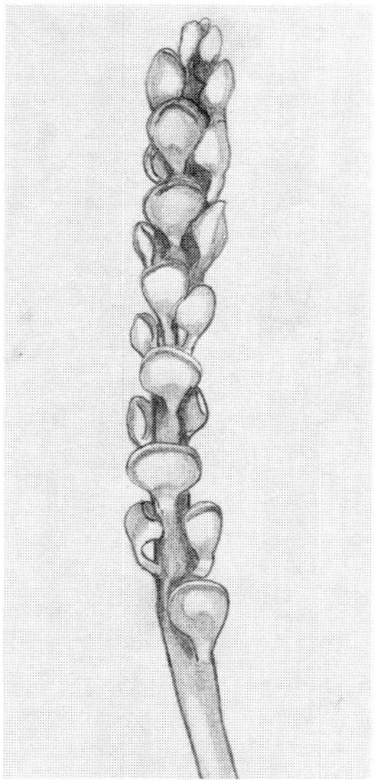

(left) Zosterophyllum *sp. showing axes with sporangia (arrowed) along their sides. Ditton Series, Brown Clee Hill, Shropshire (x 2.5)*

(right) *Reconstruction of a spike of* Zosterophyllum *based on specimens from the Ditton Series of Angus, Scotland (x 1.5)*

in which the axes were smooth, but not always regularly branching, and the sporangia were not terminal but lateral and aggregated into spikes. These new features characterise the genus *Zosterophyllum,* the earliest forms of which appear in Scotland, but by the late Ditton it was also established in the Welsh Borderland. The type of organisation seen in *Cooksonia* is still present in these floras but here the sporangia are relatively long and cigar-shaped. It is difficult to assess the affinities of these early vascular plants since no living forms have exactly similar structures but perhaps the closest are the modern Psilotales, a small group with tiny leaves, small clusters of sporangia and no roots, typified by the genus *Psilotum.*

Before tracing further developments it is now relevant to return briefly to fossils from the Lower Palaeozoic to examine additional early plant evidence in the form of spores, already mentioned above, and other microfossils.

Palynology

The study of fossil spores, pollen and microplankton which can be isolated from rocks by chemical means, provides a great deal of information on plant evolution, including supporting evidence for the early history of land plants. The word palynology itself, originally applied to the study of modern pollen and spores but subsequently also to fossils, has strong Welsh associations, being coined in 1944

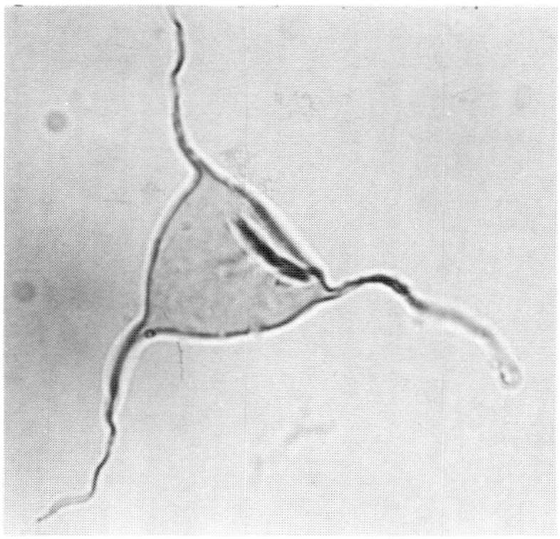 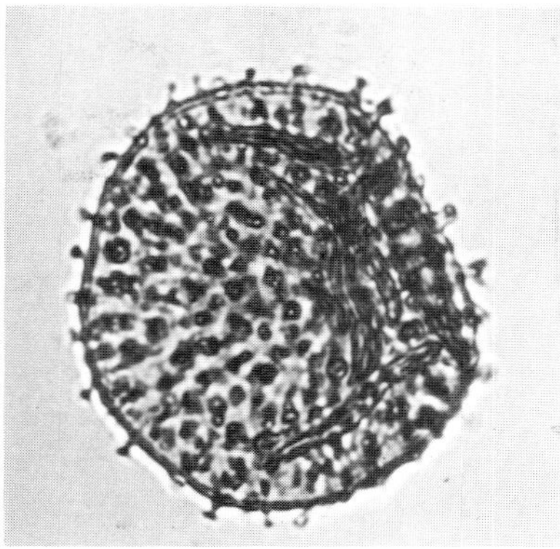 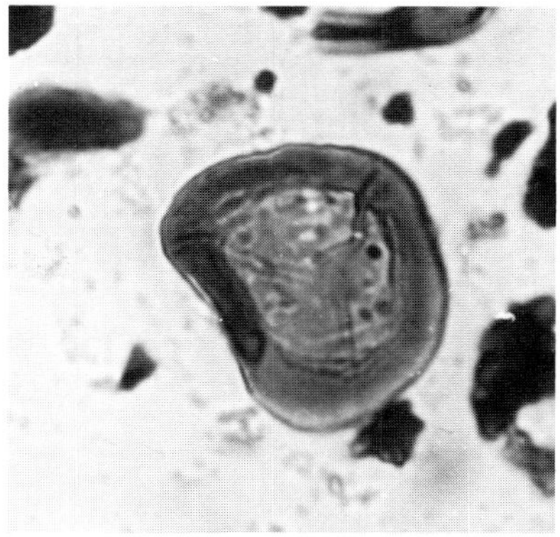

left *and* **centre** *Silurian acritarchs:* **(left)** Veryhachium *sp.*, **(centre)** Visbysphaera *sp. Downton Series, Welsh Borderland.* **(right)** *A Silurian trilete spore* Archaeozonotrilites chulus nanus, *showing typical 'Y' shaped mark. Downton Series. Welsh Borderland. (All x 1000)*

by Dr. H.A. Hyde, former Keeper of Botany at the National Museum, and Dr. D.A. Williams of Llandough Hospital, who together collected pollen on the roof of the Museum in connection with the investigation of chest diseases.

The relative abundance of microfloras in a wide variety of rock types from which other fossils may be unknown makes them particularly useful for stratigraphical dating and correlation. In Dyfed, for example, microfloras occur in the otherwise unfossiliferous Newgale Beds and provide a firm Middle Cambrian date, whereas before the floras were discovered in 1964 estimates of the age of the beds ranged from Cambrian to late Carboniferous. The Newgale fossils are mainly acritarchs, some of which, as mentioned earlier, might represent cysts of unicellular algae known as dinoflagellates, the cyst forming a protective sac around the plant. Ordovician acritarchs have been described from phosphatic Caradoc limestones around Pen-y-Garnedd in Clwyd, and rich assemblages are known from the Silurian throughout Wales and the Welsh Borderland, notably in the Wenlock and Ludlow. Since the acritarchs are abundant only in full marine sediments, they decrease rapidly in numbers across the boundary from the marine Ludlow to the brackish and non-marine Downton rocks and become insignificant or absent through the remainder of the Old Red Sandstone, before re-appearing in the marine Carboniferous.

Tetrads of fossil spores first appear in the Ordovician, the oldest yet described being from the Caradoc of North Africa, where a few simple types are present. Isolated spores with trilete marks from the Welsh area are important in tracing subsequent developments throughout the Llandovery, Wenlock, Ludlow and Downton, mainly involving a gradual increase with time in numbers and type of spores;

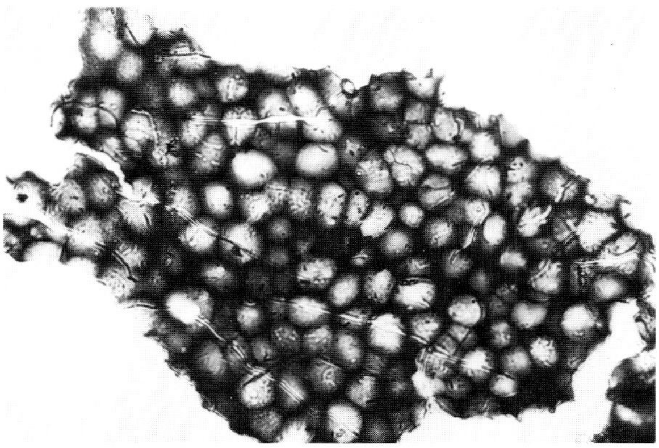

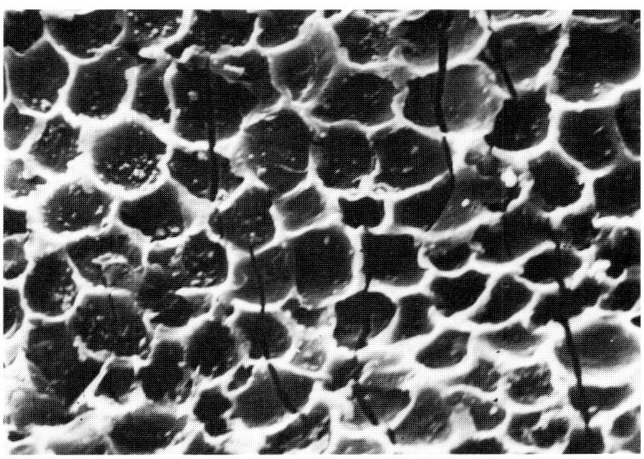

Piece of cuticle of Nematothallus*:* **(above)** *viewed under light microscope; the dark lines are thickened regions of cuticle where it penetrates between the surface cells and thus indicate their outlines (x 350).* **(below)** *scanning electron micrograph showing detail of the cells (x 800)*

these changes are generally considered to reflect early evolutionary diversification in vascular plants, although there is no direct evidence of the parent plants in pre-Wenlock rocks. Since some living plants such as mosses, liverworts and a few algae also possess trilete spores, the alternative interpretation, that Caradoc, Llandovery, Wenlock and Ludlow spores evolved independently in various early plants, before other truly vascular characters appeared, cannot be wholly discounted. Indeed, there is further evidence in the dispersed microfossil record that vascular plants were not the only colonisers of the land in Silurian times. Small fragments of cuticle bearing imprints of the outermost cellular layer of the plant, but lacking stomata, are frequently recovered when Silurian rocks are macerated. They have also been removed from irregularly shaped, coalified patches several millimetres in diameter from certain localities, e.g. Capel Horeb. It has therefore been postulated that these cuticles covered flattened (thalloid) encrusting plants which grew on the land, but in the absence of any information on their anatomy or reproductive structures the affinities of these land plants, called *Nematothallus*, remain unknown.

From Ditton times onwards, the close association of most spores with land plants is unequivocal, and the next stages of evolution can be traced in the important floras of the Old Red Sandstone.

Old Red Sandstone floras

Following the initial invasion of the land there was a very rapid increase in the number of plant genera and species and in the evolution of the various types. Terrestrial environments necessary for their development were available in several parts of the world, notably on the great northern Old Red Sandstone continent which was established by uplift and mountain building through Silurian-early Devonian times, with its southern margin close to the line of the present Bristol Channel; sediments deposited in the rivers, lakes and deltas of the continent contain most records of the plants washed from the land surface, although some survived transport to the open seas to the south.

The significance of the early *Cooksonia* and *Zosterophyllum* floras in the Downton and Ditton rocks of the Welsh area has already been outlined. Subsequent history through the Lower Old Red Sandstone is based largely on floras from the Senni Beds of the Brecon Beacons, one of the most important horizons in the world for Lower Devonian plants, and one from which new forms are still being found. Among the described species there is a variety of architecture reflecting evolutionary diversification. *Cooksonia* and *Zosterophyllum* are still present, though now represented by different species than in the Ditton beds, and one of their most common and distinctive associates is *Gosslingia breconensis*.

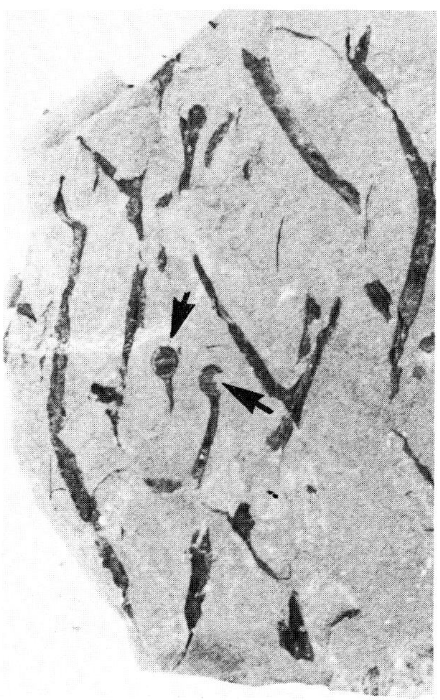

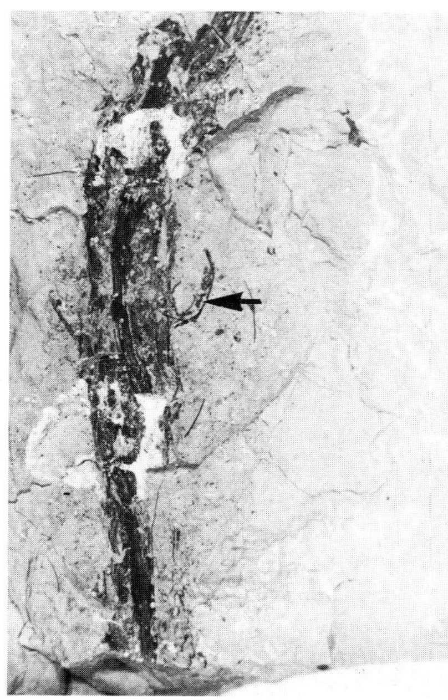

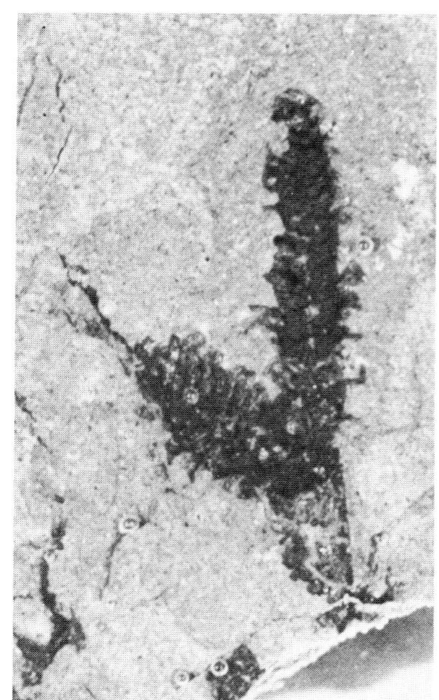

(left) Cooksonia *sp., Senni Beds, Brecon Beacons; compare with the earlier Downton specimen and note the larger axes with more prominent, oval sporangia (arrowed) (x 1).* **(centre)** Drepanophycus spinaeformis, *Senni Beds, Brecon Beacons; note the thick, dark, central vascular strand and the thorn-like leaf (arrowed); the earliest example of a leaf in Wales (x 1).* **(right)** Krithodeophyton croftii, *Senni Beds, Brecon Beacons; a forked fructification composed of rows of lateral sporangia (x 3.5 approx.)*

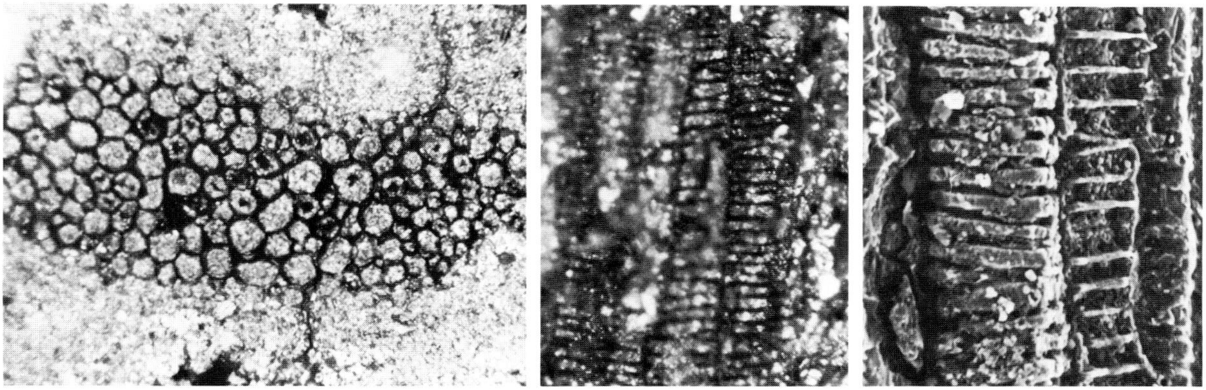

Transverse **(left** *x 80 approx.*) *and longitudinal* **(centre** *x 160 approx.*) *sections through the centre of a petrified axis of* Gosslingia *in the region of the water conducting tissue (xylem composed of tracheids).* **(right)** *Scanning electron micrograph showing two tracheids; the right hand one shows the lignified rings which prevent the collapse of the tubes under tension; in the left hand tracheid the centre of the tube is filled with pyrite and has moulds of the thickenings on the surface (x 400 approx.) Senni Beds, Brecon Beacons*

Gosslingia is closely related to *Zosterophyllum*, having smooth, branching axes and lateral sporangia, but the sporangia are more scattered, tending to be arranged in groups as rows along one side of an axis. *Dawsonites* and *Drepanophycus* are also common associates, the former representing a possible forerunner of a complex of plants in the Middle Devonian from which ferns and gymnosperms probably evolved, while the latter genus is a small herbaceous form resembling modern lycopods such as the club-moss *Lycopodium*; early descendants of *Drepanophycus* may have given rise to some of the massive trees in the Carboniferous, but the form of the modern club-moss is more similar to the early habit of this group. *Krithodeophyton* is another distinctive Senni Beds plant, with spikes of sporangia which are arranged in two rows, one on either side of the axis, alternating with slender, sterile appendages attached at right angles to the axis; its exact relationships are still uncertain.

The value of the Senni Beds floras is enhanced by the preservation of some specimens in iron pyrites as three dimensional petrifactions, from which thin sections can be cut and examined microscopically to determine fine structural details. Similar floras to those in the Brecon Beacons are now known from a number of localities in south-east Dyfed, and their description in the near future should add considerable new data to details of this phase of development.

Middle Old Red Sandstone deposits are absent throughout much of the Welsh area and floras from the Upper beds are rare or absent, the two factors combining to leave extensive gaps in the local evolutionary story. Elsewhere through the Middle and Upper Devonian the most important events included the origin of the horsetails and ferns, the origin of the arborescent or tree habit, the origin of the first seed, and the prominent development of the progymnosperms as the possible ancestors of the gymnosperms or naked-seed plants.

The evolution of the parent plants through the Devonian is of course reflected in the spores, both by a continued increase in total numbers and in morphological variations, and also by the development of spores of two distinct sizes. Plants bearing one type of sporangium that contains spores of uniform size are called homosporous, and all early Devonian plants are thought to have been of this type. By Middle Devonian times some spores exceeded 200 microns in diameter, an arbitrary figure for the demarcation of microspores and megaspores, the combination of both types in different sporangia on the parent plant being referred to as the heterosporous condition; the megaspores produce the female sexual organs. This step appears to have been consolidated during the Middle Devonian and eventually resulted in the seed habit.

The earliest known seeds are from the Upper Devonian of North America; a flora from the highest Old Red Sandstone near Tongwynlais, north of Cardiff, contains the first indications of British forms, some identical to the American material and others with an extremely simple structure unlike anything described from elsewhere. In this flora the seeds are accompanied by other plant fragments referred to *Xenotheca* and *Platyphyllum*, and by a rich spore assemblage dominated by species of *Hymenozonotrilites*; the spores are very useful in correlating the highest Old Red Sandstone with marine sequences of Europe and suggest a low Tournaisian age. Fragmentary plants of similar age occur in the Skrinkle Sandstone at the top of the Old

(left) *A stalked seed (x 4.5);* **(right)** *a group of sporangia containing microspores, similar to* Xenotheca *sp. (x 2.3); Upper Old Red Sandstone, Tongwynlais, near Cardiff*

Red Sandstone in Dyfed.

The importance of vascular plants throughout the Devonian overshadows contemporaneous changes in the algae, mainly involving the extinction of older Palaeozoic types such as *Prototaxites*, and the establishment of new ones that became common in Carboniferous and later times; however, while many of the massive calcareous rock builders continued to exist, they are not found in the terrestrial sediments of the Old Red Sandstone. One new group which originated in the late Silurian was the charophytes, represented today by the stoneworts or brittleworts; they are found in fresh or slightly brackish water environments and occur rarely in the Upper Old Sandstone of South Wales.

Carboniferous Limestone and Millstone Grit floras

The erosion and lowering of the Old Red Sandstone continent finally led to the re-incursion of the sea over much of Britain at the beginning of the Carboniferous Period (345 million years ago),

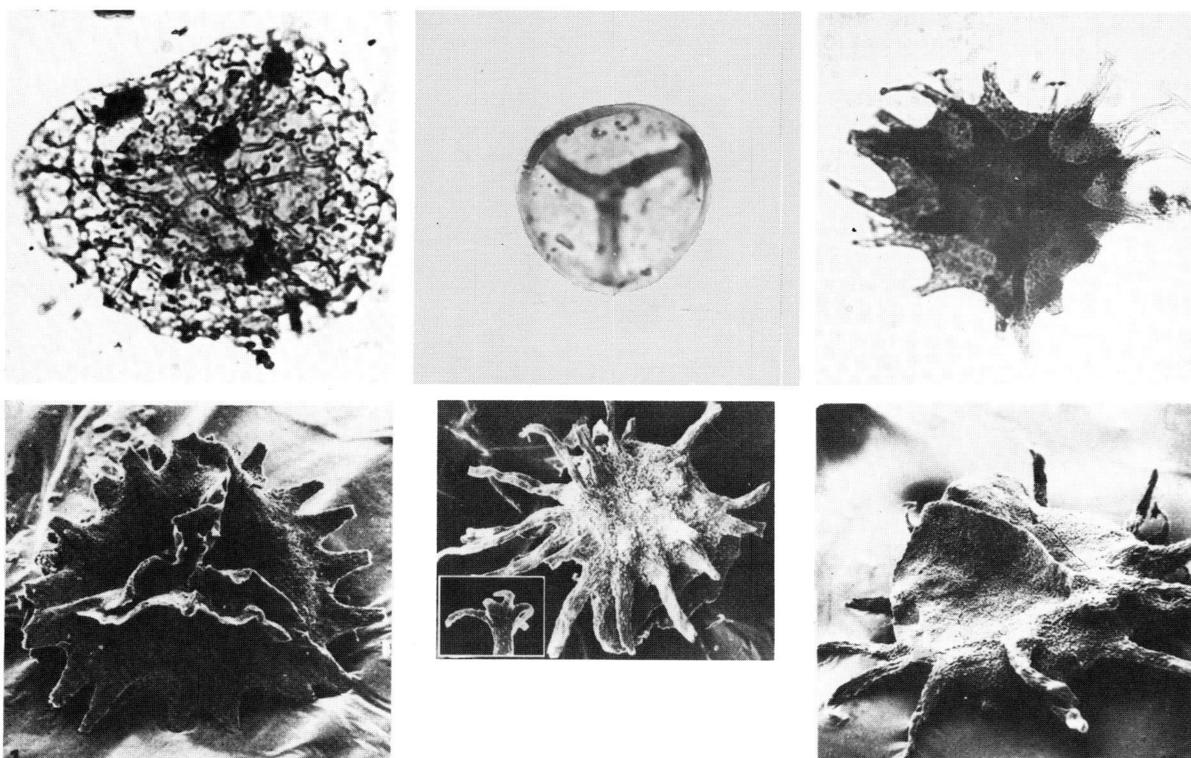

Spores from the Upper Old Red Sandstone of Tongwynlais, near Cardiff, showing a wide variation in morphology. **(top left)** Hymenozonotrilites lepidophytus *(x 500);* **(top centre)** Retusotrilites incohatus *(x 500);* **(top right)** Ancyrospora capillata *(x 200). The bottom three photographs are scanning electron micrographs showing three-dimensional features of the spores.* **(bottom left)** Ancyrospora capillata, *a megaspore with well developed triradiate mark (x 180);* **(bottom centre)** Hystricosporites cf. delectabilis – *inset shows the multifurcate tips of a spine (x 126);* **(bottom right)** Hystricosporites multifurcatus *megaspore (x 140); (SEM photographs by permission of J. Robson and the editors of* New Phytologist)

resulting in the widespread deposition of the Carboniferous Limestone. Where these deposits now crop out across North and South Wales they provide renewed evidence of the role of calcareous algae in rock building. Older Palaeozoic genera such as *Girvanella, Ortonella, Rothpletzella,* and *Solenopora* are again important, joined by new forms such as *Garwoodia,* probably closely related to *Girvanella*, and by the dasycladacean *Koninckopora*. Small reefs and stromatolitic mounds in the Llanelly Formation, within the Carboniferous Limestone sequence, contain some of these genera, together with further laminar forms referred to *Bevocastria, Mitcheldeania* and *Stylocodium*. The algal laminates often form encrusting layers or concentric growths around pisoids, or occur as oncolites. The occurrence in the algal laminae of a variety of minute tubes (porostromates) similar to Recent calcified cyanophytes suggests that the laminae were built partly by filamentous cyanophytes. Sedimentary evidence indicates that at this time environmental conditions varied from normal marine, through highly saline, to freshwater, and it has been suggested that some of the stromatolites may have formed as subaerial crusts. The presence of new marine algal genera throughout the Carboniferous

Stromatolites in Carboniferous Limestone, Llanelly Formation, Blaen Onneu, near Crickhowell, Powys. Note the fine, gently domed algal lamination (x 1 approx.)

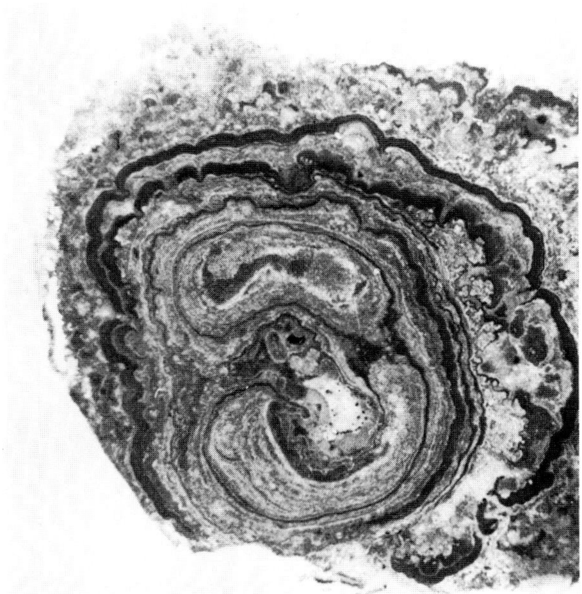

Polished hand specimen of pisoid with distinctive algal layering (the light coloured laminae are algae). The algae grew as successive coatings around a core of aragonite, probably in a hypersaline environment. Carboniferous Limestone, Llanelly Formation, Blaen Onneu, near Crickhowell, Powys (x 1.5 approx.)

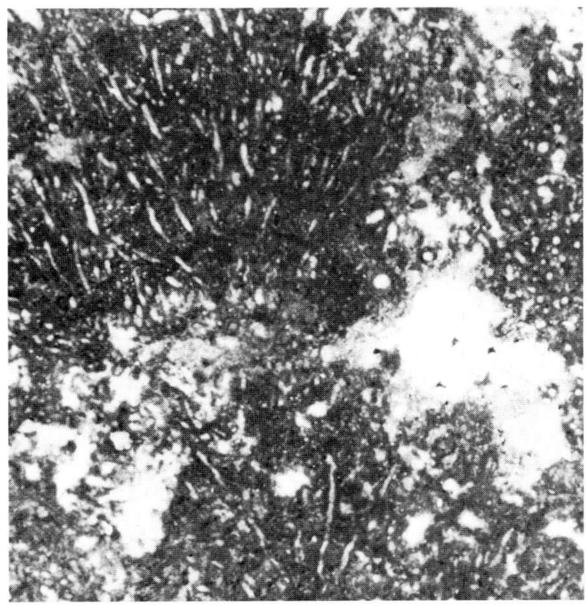

Thin sections of calcareous algae from the Carboniferous Limestone. **(above)** Garwoodia *sp., an oncolite (oncoid) showing radial growth; Llanelly Formation, Blaen Onneu, near Crickhowell, Powys (x 8 approx).* **(below)** Ortonella *sp., a bush of algal tubes; Llanelly Formation, Cwar yr Ystrad, near Trefil, Tredegar, Gwent (x 8 approx.)*

Limestone reflects the diversification that this group had undergone during Devonian times in other parts of the world.

Records of land plant remains in the Carboniferous Limestone are much less common, but in the more sandy and muddy rocks deposited fairly close to the coastline of islands and larger masses which stood out in the sea, drifted vegetation is sometimes sufficiently well preserved to allow the main developments of the land floras to be followed. One such flora occurs across much of North Wales in shale beds within the Lower Brown Limestone division of the Carboniferous Limestone, which was deposited close to the northern margin of a land mass spreading across central Wales and parts of the English Midlands. The most interesting aspect of this flora is the presence of abundant and varied arborescent lycopods, including *Lepidodendron, Lepidodendropsis, Archaeosigillaria* and *Stigmaria.* The earliest record of horsetails (Sphenopsida) in Wales is from the same flora, in the form of small, longitudinally ribbed stems belonging to *Archaeocalamites,* together with cones of *Bowmanites.* The fern-like foliage of *Rhacopteris* and a non-calcareous algal form referred to *Buthotrephis* are further associated plants. Another important flora from North Wales occurs in slightly younger Carboniferous Limestone at Teila, near Prestatyn in Clwyd; the major groups found in the Lower Brown Limestone are again present, together with a more diverse assemblage of seed-ferns, of which *Diplopteridium* is the most common. The plants at both these horizons present early evidence of the build-up of floras which were to become so dominant in the late Carboniferous, and during Namurian times the picture was essentially one of an acceleration of the build-up as the vegetation spread across the deltaic environments produced by the accumulation of sediments eroded from the land masses and the consequent silting up of the Carboniferous Limestone sea.

The Coal Measures forests

By 300-310 million years ago the deltas had spread southwards over much of Britain to produce a low-lying terrestrial area covered largely by fresh-water swamps, marshes and lagoons, separated by slightly elevated tracts of drier land. Similar conditions were widespread throughout the northern hemisphere, and in those regions which lay in tropical and sub-tropical latitudes, as did Britain at that time, the climate allowed the rapid growth of luxuriant forest vegetation. Great thicknesses of sediments accumulated during prolonged gentle subsidence of the earth's crust, including layers of peat which were formed when plants died and fell into the swamps;

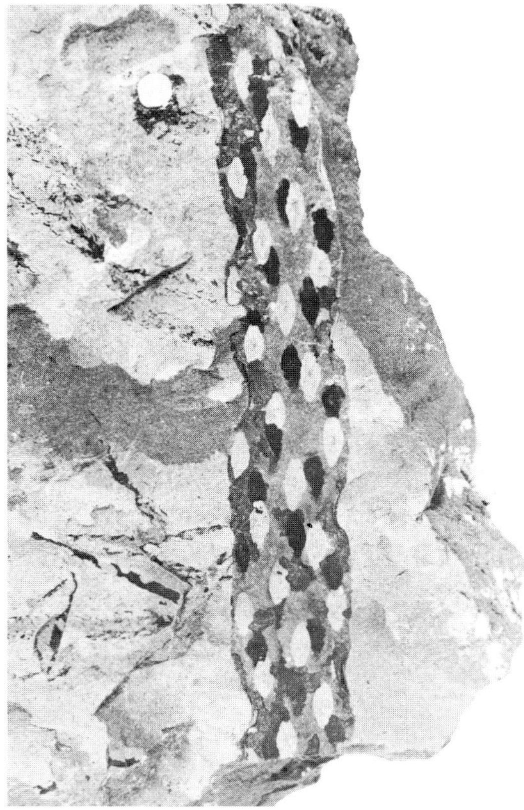

Defoliated branch of Lepidodendron perforatum. *Lower Brown Limestone (Carboniferous Limestone), Graig Quarry, Clwyd (x 0.9 approx.)*

Archaeosigillaria [Clwydia] decussata *showing leafy shoot tips. Lower Brown Limestone (Carboniferous Limestone), Graig Quarry, Clwyd (x 1)*

Reconstruction of Coal Measures forest. L = Lepidodendron*; S =* Sigillaria*; P = a seed fern, note the seeds attached to ends of fronds; F = a tree fern; C =* Lepidostrobus *(cones); H =* Sphenophyllum*, an herbaceous horsetail; M = lycopod leaves. Photograph by permission of the Field Museum of Natural History, Chicago*

the subsidence repeatedly led to the deepening of water which destroyed the living vegetation, and on a number of occasions allowed the sea to flood over the area. Sediments were then brought into the deepened water, leading to progressive shallowing until the area again became fully terrestrial and the forests were re-established. The layers of peat were converted to coal by compaction by the overlying sediments, and repetitions of the complete cycle produced the sequence of coal seams now seen in the coalfields such as those of North and South Wales.

The density of vegetation in the Coal Measures forests is best matched today by that of tropical rain forests. It has been estimated that a mass of plant debris as much as 4 to 6 metres thick was necessary to form a 0.5 metre coal seam. In this great compaction to form masses of coal the structure of individual plants was generally destroyed, with the result that the best preserved specimens of the period are now found in the sediments between the seams themselves. Although some of these plants may have grown on the peat bogs, the majority probably lived outside the swamp environment and were washed in during times of flooding.

One of the most striking features of Coal Measures plants was the occurrence of large, tree-sized forms in groups which are now represented only by small herbaceous types. The most spectacular examples are in the lycopods and horsetails. Such large plants invariably broke up on transportation and burial or were only partly preserved, resulting in very fragmentary fossils consisting of discrete pieces of stem, roots, leaves, seeds or sporangia. As with fossil plants of all geological periods, these individual organs that once made up a parent plant are given separate names; for example, in one arborescent lycopod the stem is called *Lepidodendron*, the root system *Stigmaria*, the leaf *Lepidophyllum* or *Lepidophylloides*, and the cone *Lepidostrobus*. The rare specimens in which these are found still joined

Terminal branches of the tree lycopod Lepidodendron acutum *with attached leaves* (Lepidophyllum) *and pendulous cones* (Lepidostrobus, *arrowed) on the branch tips. These organs are seldom found together on a single specimen, making this a rare and valuable fossil. Coal Measures, shales associated with Pentre Seam, Gilfach Goch, Glamorgan (x 0.5 approx.)*

together indicate that they were so related.

Much of the foliage of Coal Measures plants bears a superficial resemblance to modern ferns, to the extent that in the past the whole period was often called the 'age of ferns'. Whereas the true ferns reproduce by means of spores, it is now known that the most common Carboniferous types were fern-like plants which bore seeds. In the absence of reproductive parts it is difficult to distinguish between the two groups using other characters such as the form of the fronds; in fact, since the overall shape of the leaf is sometimes very similar in both groups, and since this criterion is one of those used in identification, some 'genera' such as *Sphenopteris* may include both true ferns and seed-ferns. Undoubted ferns were present in the coal forests, some of them reaching heights of over 15 metres; their foliage is called *Pecopteris* and more rarely *Sphenopteris*. *Asterotheca* is one of the names given to compression fossils of fertile fronds; its pinnules have

Pecopteris plumosadentata; *a true fern with sterile (left) and fertile pinnae. Coal Measures, Rhigos, near Hirwaun, Mid Glamorgan (by permission of the Trustees, British Museum (Natural History), Taylor Collection) (x 0.9)*

Eusphenopteris obtusiloba; *note that the leaflets (pinnules) are lobed. Coal Measures, Rhigos, near Hirwaun, Mid Glamorgan (by permission of the Trustees, British Museum (Natural History), Taylor Collection) (x 0.6)*

a mottled appearance, which indicates the presence of numerous clusters of squashed sporangia on the lower surface.

The seed-ferns or pteridosperms include a wide variety of well known leaf genera such as *Neuropteris, Alethopteris, Sphenopteris, Eusphenopteris, Mariopteris, Lonchopteris, Cyclopteris* and *Odontopteris;* the generic names are applied to whole fronds in some cases and to individual leaflet types in others. Differences between the genera are based on numerous characters such as the overall shape and branching of the fronds, the shape of the leaflets (pinnules) together with the details of vein structures, and the position and nature of the attachment of the pinnules to the leaf stalks.

The pteridosperms belong to a major line of gymnosperms whose living representatives are the cycads, many of which are palm-like trees with a crown of large fronds. It seems very likely that the Palaeozoic pteridosperms were the ancestors of both the cycads and the third group in this gymnosperm

Neuropteris loshi?, *a frond composed of numerous leaflets (pinnules) showing the diagnostic characters of the form genus* Neuropteris; *the arrow indicates a seed. Coal Measures, Racey Seam, Danygraig, near Swansea, West Glamorgan (x 0.5 approx.)*

Neuropteris obliqua, *Coal Measures, Rhigos, near Hirwaun, Mid Glamorgan (by permission of the Trustees, British Museum (Natural History), Taylor Collection) (x 0.8)*

Cyclopteris orbicularis; *this type of pinnule is found attached to the basal regions of a* Neuropteris *frond. Coal Measures, No.2 Rhondda Seam, Ynyshir, Rhondda, Mid Glamorgan (x 1)*

Alethopteris lesquereuxi, *a beautifully preserved frond fragment showing the central midribs and extended bases of the leaflets. Coal Measures, Forest of Dean, Gloucestershire (x 1)*

Diplothmema *sp., a portion of a frond in which the leaflets are deeply divided. Coal Measures, above 9ft Seam, Cymmer, Mid Glamorgan (x 1)*

Mariopteris nervosa, *Coal Measures, Rhigos, near Hirwaun, Mid Glamorgan (by permission of the Trustees, British Museum (Natural History), Taylor Collection).* Mariopteris *is characterised by a bifurcation at the base of the frond (x 1.5 approx.)*

Reconstruction of Medullosa noei. *The leaves belong to the form genus* Neuropteris *and the stem with its 'supporting' roots is called* Medullosa; *the complete plant was apparently 5 metres high (by permission of Prof. W.N. Stewart and the editors of* Botanical Review*)*

Part of a long, strap-shaped leaf of Cordaites principalis. *Coal Measures, Limog Vein, Wernddu, Caerphilly, Mid Glamorgan (x 1 approx.)*

line, the bennettitalians, both of which appeared in some numbers in the Mesozoic.

Two further types of seed-bearing plants are recorded in the late Carboniferous, and in contrast to the pteridosperms the arborescent habit was strongly developed, with trunks possessing abundant secondary wood composed of tracheids. Both belong to the second major line of gymnosperms represented in modern floras by the conifers (pines, firs, redwoods). The conifers themselves are not found commonly in Coal Measure floras; they appear to have originated in the late Carboniferous, becoming increasingly diversified in succeeding geological periods. Closely related are the Cordaitales, now extinct, but represented most commonly in the Carboniferous by *Cordaites* which possessed long strap- or sword-shaped leaves with fine, parallel veins. Intact plants have never been found, but it is apparent from the sizes of some of the pieces of fossil wood that the trees were tall, some over 40 metres high with leaves borne on a crown of branches. Unlike present day conifers the seeds of cordaites were not borne in compact cones, but were carried together with small leaves on short, catkin-like branches.

Undoubtedly the most spectacular plants of the coal forests were the giant lycopods, which attained heights of 40 metres or more in well known tree forms such as *Lepidodendron* and *Sigillaria*; the trunks of these trees were thick and straight, with the branches formed close to the top. In *Lepidodendron* the terminal branches bore spirally arranged,

(left) Lepidodendron aculeatum; *part of the trunk showing typical diamond-shaped leaf bases. Coal Measures, Burslem or Llay Main Pit, near Wrexham, Clwyd (x 1).* **(right)** Lepidodendron *cf.* lanceolatum; *note the difference in shape and size of the leaf bases from the specimen on the left. Coal Measures, 7 ft Seam, Cymmer Colliery, Porth, Mid Glamorgan (x 0.8 approx.)*

elongated, pointed leaves, while the main trunk was covered with distinctive diamond shaped markings which represent the shapes of the leaf bases or cushions, the leaves themselves having been shed. The patterns of these leaf cushions are important in distinguishing different genera and species; in *Sigillaria* they are normally hexagonal in shape and arranged in vertical rows, while in another common arborescent genus, *Bothrodendron,* there are no cushions and the leaf scars are small and circular. *Lepidophloios* is a close relative of *Lepidodendron,* but in the latter genus the diamond shaped scars are elongated vertically whereas in the former they are typically transverse. In addition to the regularly arranged leaf cushions, some stems occasionally bear large circular scars indicating the attachment of cast off branches; these are known as ulodendroid scars.

The lycopods, like the ferns, bear sporangia which contain spores, but whereas in the majority of ferns numerous sporangia are attached to the lower surface of a frond, in the lycopods each fertile leaf (sporophyll) bears a single sporangium near its base and the sporophylls themselves are aggregated into cones. In *Lepidodendron* such cones, named *Lepidostrobus,* hung from the ends of the terminal branches.

The underground parts of many of the arborescent lycopods are so similar that they are included in the same genus, *Stigmaria.* These thick, branching, root-like structures occur most commonly in the seat-earths or underclays immediately below coal seams. The recognition of seat-earths as fossil soils in which the rich vegetation grew was made in the early 19th century, and many of the details of their nature and significance resulted from the work in South Wales in 1840 of Sir W.E. Logan, later Director of the Geological Survey of Canada. In 1838 Logan found a number of large trunks of *Sigillaria* in position of growth in rocks forming the steep side of Nant Llech in the Swansea Valley; two of the trunks were excavated and removed to the Royal Institution at Swansea, under the supervision of Sir H.T. de la Beche, first Director of the Geological Survey of

Large branch scar of ulodendroid type. Coal Measures, Nant Melin, Aberdare, Mid Glamorgan (x 0.75 approx.)

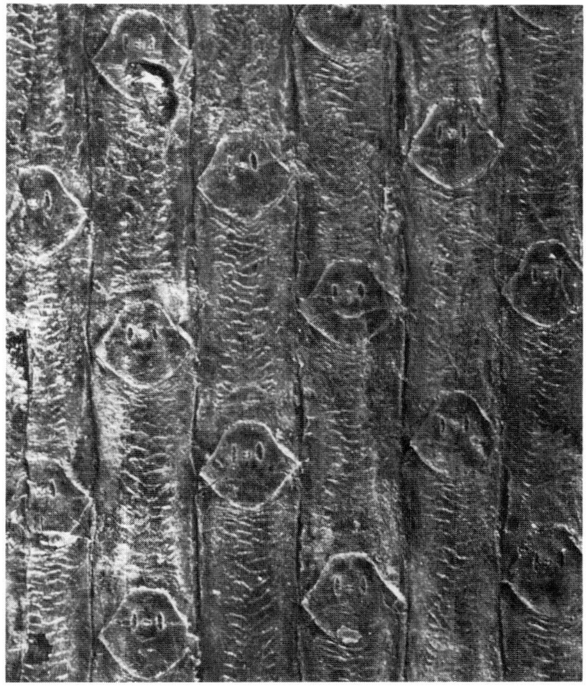

Sigillaria scutellata; *portion of the trunk showing leaf bases arranged in vertical rows. Coal Measures, South Wales (x 1.5)*

Great Britain. A contemporary account of the discovery suggested that 'it is not extravagant to imagine, that were the sides of the dell which cut it cleared away, a whole primeval forest of those gigantic *Segillaria,* standing as they grew, would be exhibited to the wondering eyes of the beholder.' Similar fossil forests have since been excavated elsewhere to reveal just these features, the best example in Britain being that at Victoria Park in Glasgow. Modern methods of coal mining often destroy spectacular specimens such as tree trunks, although many are still encountered in both open-cast and underground workings. One of the best examples found in Wales in recent years was discovered in February 1982 at Six Bells Colliery, Abertillery, Gwent, rooted vertically in a 0.3 metre thick seat-earth about 1 metre above the Yard Seven Foot Meadow Vein; it was found at a depth of 405 metres while handworking a drive from Six Bells to Blaenserchan, Pontypool, and since the seam was not being worked by machinery it was possible to remove the trunk more or less intact. It is a specimen of *Sigillaria* some 2 metres high, almost 1 metre in diameter at its widest point, and with traces of a *Stigmaria* root system attached.

Not all the Coal Measure lycopods were tall trees, but herbaceous members may be difficult to recognise because of their similarities to the terminal

Fossil tree-trunks in position of growth at Nant Llech in the Swansea Valley. From a 19th century illustration

(above) Lepidostrobus ornatus; *a cone, cut open to reveal in the basal region the central axis with lateral leaves bearing sporangia (sporophylls); the tapering ends of the leaves can be seen along the margins of the cone. Coal Measures, Abercarn, Gwent (x 1)*

(right) Sigillariostrobus nobilis; *these cones are thought to have been attached to* Sigillaria; *they were either completely microsporangiate or completely megasporangiate. Coal Measures, Rhigos, near Hirwaun, Mid Glamorgan (by permission of the Trustees, British Museum (Natural History), Taylor Collection) (x 0.75 approx.)*

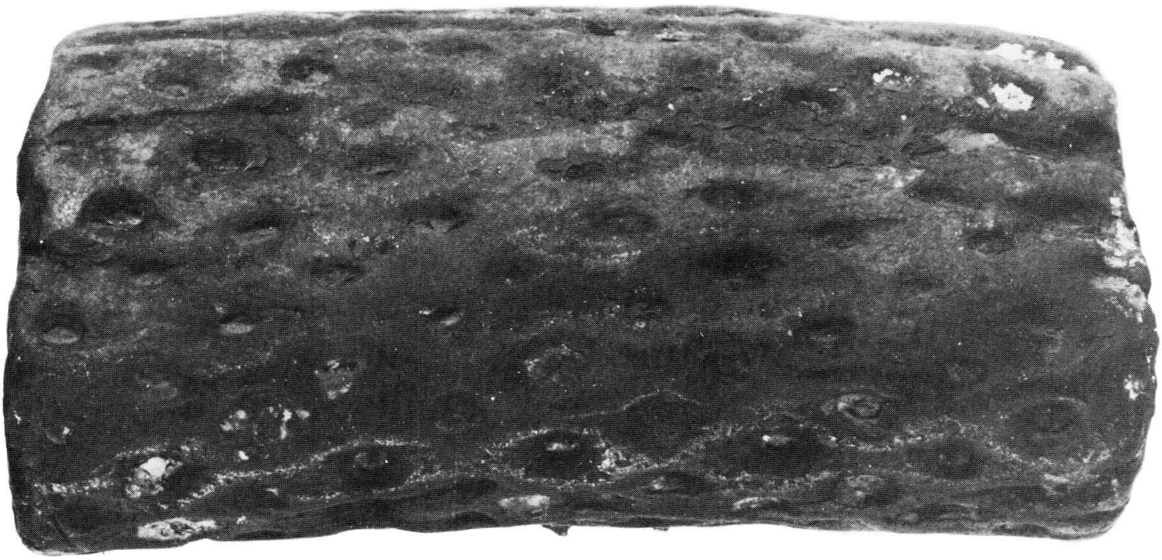

Stigmaria ficoides; *part of the underground branching system of a tree lycopod; the circular areas indicate the points of attachment of the rootlets. Coal Measures, New Tredegar, Gwent (x 1)*

Stigmaria ficoides; *axis with attached rootlets. Coal Measures, Ffaldau Colliery, Pontycymmer, Mid Glamorgan (x 0.6)*

Stigmaria ficoides; *cast of an axis showing remains of internal structure. Coal Measures, South Wales (x 0.75 approx.)*

leafy twigs of the arborescent forms. However, there are records of plants similar to modern *Selaginella* which may have formed part of the undergrowth in the forests, and it has also been suggested that such plants may have contributed to a heath vegetation.

The final important plant group in the coal forests was the horsetails, represented today by the common *Equisetum.* Large trees were again developed, the most common being *Calamites,* which had a straight, smooth stem with lateral branches in opposite pairs or whorls; in some specimens whorls of narrow, pointed leaves named *Annularia* and *Asterophyllites* can be seen to originate at nodes along the branches. *Calamites* carried its spores in cones attached to slender branches. Herbaceous horsetails were also common, restricted to a probable maximum height of about 1 metre. *Sphenophyllum* is an example of a herbaceous horsetail which probably had a scrambling habit. Narrow stems bore whorls of wedge-shaped or more finely divided leaves, some of which ended in recurved spines. The growing plants must have closely resembled the modern larger bedstraws or goose-grass, supporting themselves by hooking on to other vegetation. Cones of the herbaceous types are included in the genus *Bowmanites.*

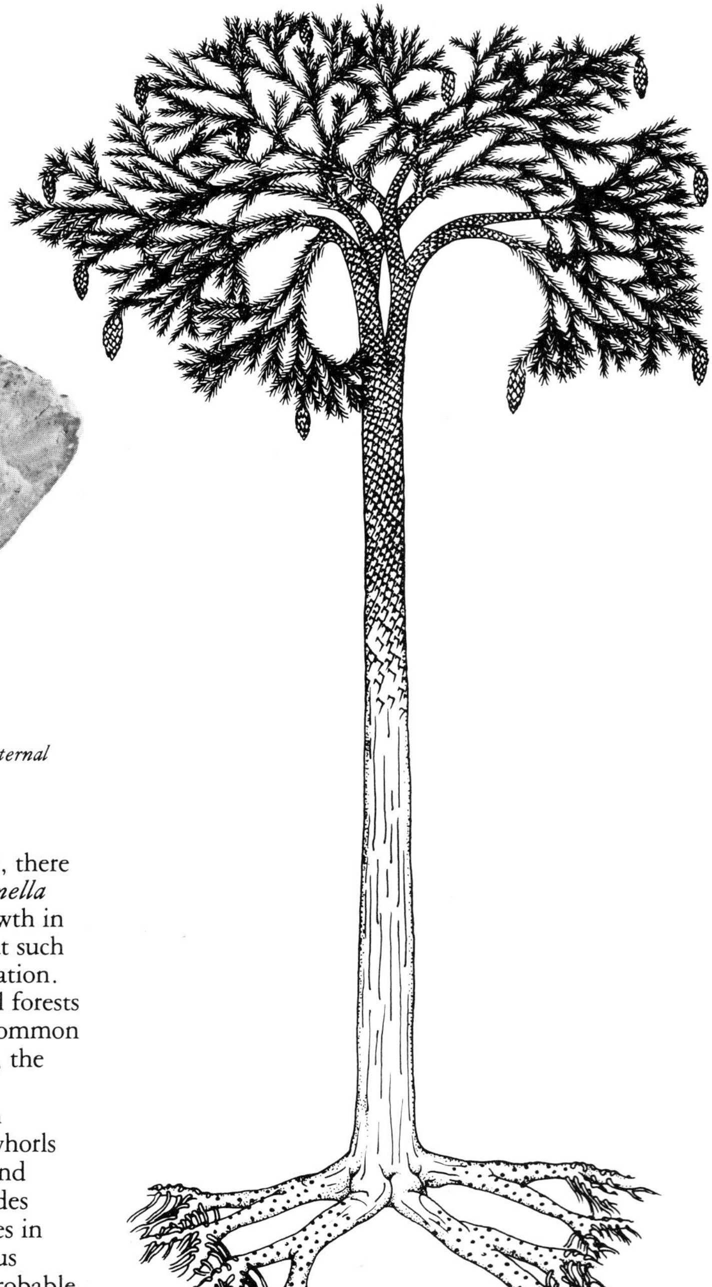

Reconstruction of Lepidodendron; *unbranched trunks over 30 metres long have been recorded for some species (by permission of Prof. D. Eggert)*

Well preserved trunk of Sigillaria, *Six Bells Colliery, Abertillery, Gwent; traces of the root system are visible near the base, which is bedded in a seat-earth above a thick sandstone sequence (by permission of Mr. J. Thomas and the National Coal Board, South Wales Area)*

Calamites approximatus; *the centre of a* Calamites *stem was hollow and frequently became filled with sediment on burial. The surrounding plant tissues then decayed, leaving a cast of the central region or pith; this specimen is such a cast. The encircling constrictions mark the positions of the nodes. Coal Measures, South Wales (x 0.5 approx.)*

Annularia radiata; *like* Asterophyllites *this is a leaf form of* Calamites. *Whereas the leaves in* Asterophyllites *are needle-like, here they are broader; both forms occurred in whorls at the nodes. Coal Measures, Rhigos, near Hirwaun, Mid Glamorgan (by permission of the Trustees, British Museum (Natural History), Taylor Collection) (x 1.8 approx.)*

Asterophyllites equisetiformis; *leaves of* Calamites. *Coal Measures, Rhigos, near Hirwaun, Mid Glamorgan (by permission of the Trustees, British Museum (Natural History), Taylor Collection) (x 1)*

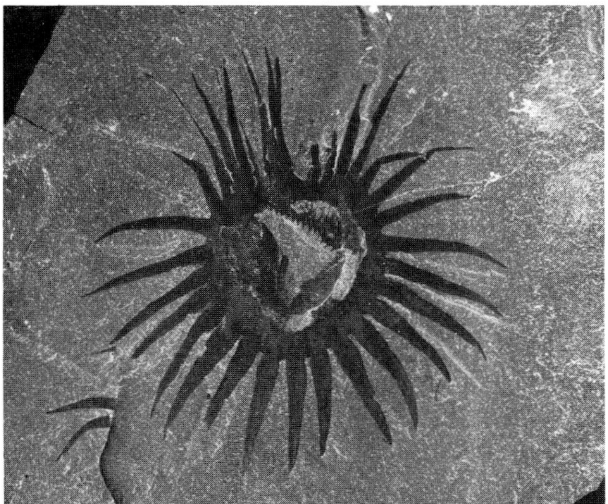

Calamites *sp., an isolated node with a whorl of leaves; the centre of the specimen would have been occupied by the stem (by permission of the Trustees, British Museum (Natural History), Taylor Collection) (x 0.8)*

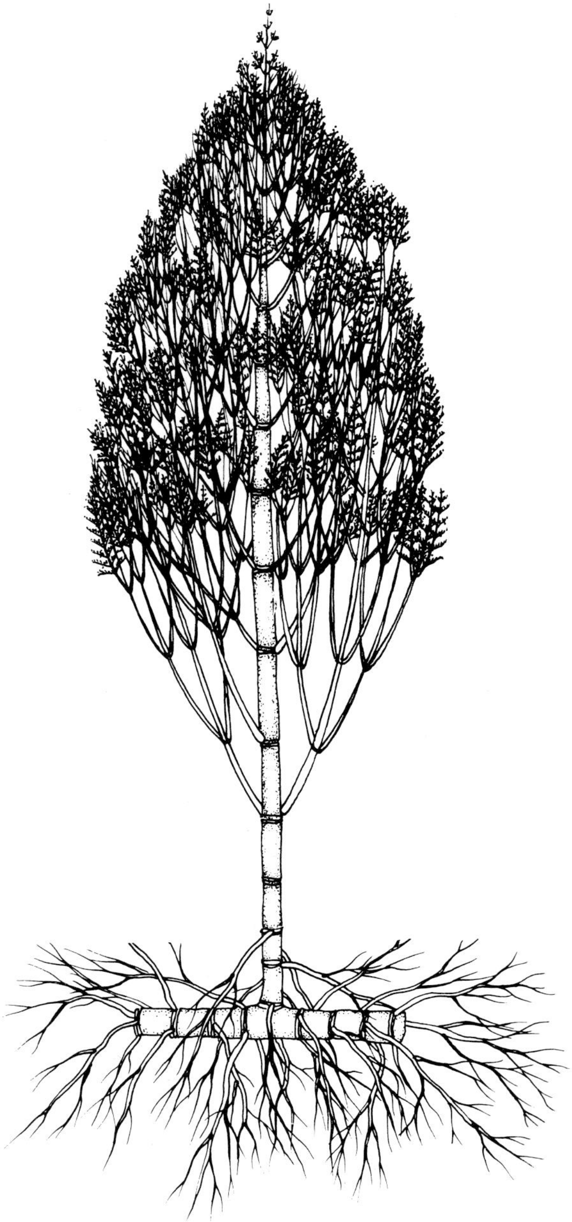

Reconstruction of Calamites (Diplocalamites) *(by permission of Prof. D. Eggert)*

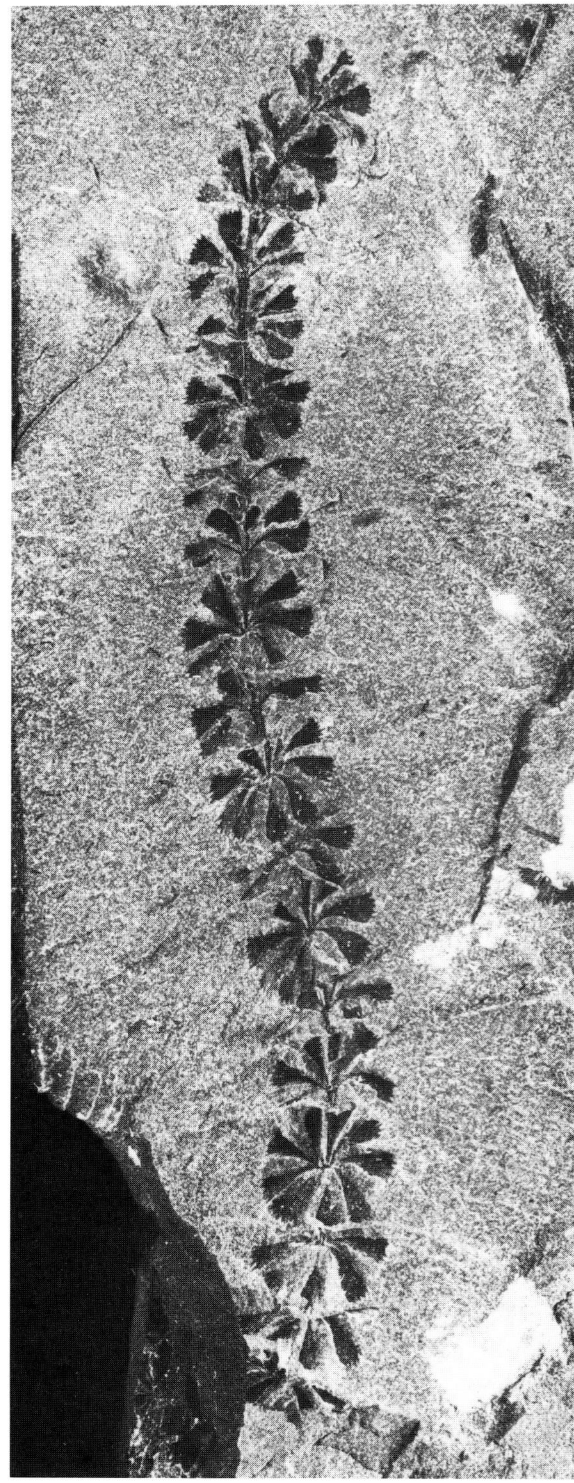

Sphenophyllum cuniefolium; *stem bearing whorls of wedge-shaped leaves with serrated margins. Coal Measures, Rhigos, near Hirwaun, Mid Glamorgan (by permission of the Trustees), British Museum (Natural History), Taylor Collection) (x 1.25)*

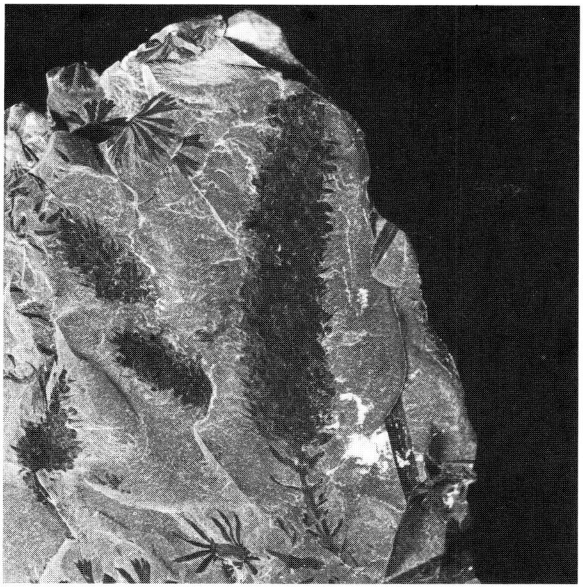

Sphenophyllum cuniefolium; *a rare specimen with attached cone* (Bowmanites). *Coal Measures, Rhigos, near Hirwaun, Mid Glamorgan (by permission of the Trustees, British Museum (Natural History), Taylor Collection (x 0.9 approx.)*

Reconstruction of Sphenophyllum emarginatum *based on specimens from the Upper Carboniferous (Westphalian D) of Germany. Note the different shaped leaves borne on the same plant, and that some of the leaves end in hooks (by permission of L.H. Batenburg and the editors of* Review of Palaeobotany and Palynology*) (x 0.4 approx.)*

Reconstruction of Coal Measures ecology

The above account gives an indication of the organisation, affinities, and diversity of vascular plants in late Carboniferous times. Of equal interest are attempts to reconstruct the ecological relationships of the floras by investigating the different associations of plants and linking these with sedimentological evidence. This approach is not new; it was first tried in detail by a remarkable pioneer in the serious study of Coal Measures plants in Wales, David Davies of Gilfach Goch in Glamorgan. Born in 1870, he began his working life as door boy in a colliery, and at an early age displayed an interest in fossils that earned him the nickname 'Dafydd Ffossil'. He eventually became a colliery agent and as his interest in geology grew he determined to study the fossils associated with the various coal seams. Davies thus began collecting fossil plants seriously in 1912 with the aim of plotting differences in the vegetation as a guide to the climate and geography in Coal Measures times, and as a method of correlating seams in different parts of the coalfield. Working mainly in his spare time he made a vast collection, sometimes breaking up 10 to 12 tram loads of rock from each seam and extracting every specimen, and he methodically analysed the number of species present, the number of specimens of each, and the type and thickness of sediments separating the seams.

From this information Davies was able to reconstruct different environments in the coal forests of South Wales. He discovered that certain plants were seldom found together and that some were associated with a distinct type of sediment, which led him to suggest that at times the area was a waterlogged swamp in which *Lepidodendron* and *Sigillaria* were dominant, while at other times it was marshy and overgrown with *Calamites,* or dry with pteridosperms such as *Alethopteris* and *Mariopteris* as the most common types.

Before his death in 1931 Davies published a number of important papers on his work, and the vast collections on which they were based are now deposited in the National Museum. Much of the stratigraphical work in zoning and correlating the Coal Measures in South Wales, notably by Drs. R. Kidston, E Dix and R. Crookall, was based in no small part on these collections which Davies made readily available to other palaeontologists.

Since Davies published his results, other workers have adopted different methods in attempts to discover more about the composition of vegetation in the diverse environments of the late Carboniferous. Their conclusions are broadly the same as his. One method involves an analysis of the spores and pollen grains which can be isolated from coals and the sediments between the seams. Such studies were initiated in the 1930s with the intention of evaluating the use of spores in correlating coal seams. The original investigations centred on megaspores, but soon switched to spores less than 200 microns in diameter, which are always far more abundant. Distinctive assemblages can often be recognised in individual seams and correlated over wide areas, and the accuracy of the results is being improved by the statistical analysis of the assemblages, together with studies to evaluate the effects of environmental control. Both the North and South Wales coalfields are yielding valuable data for these studies.

In correlating coal seams it is the composition of assemblages that is important, and there is no need in such studies to know anything about the plants which produced the spores. Palynologists introduced the term 'miospore' to encompass all types less than 200 microns in diameter, regardless of whether they came from homosporous plants, were the microspores of heterosporous plants, or were the pollen grains of seed plants. For palaeocological studies it is obviously essential to discover the parents, especially as it soon became apparent that certain types of spore (e.g. *Lycospora, Densosporites*) far outnumbered all others in most of the coals examined and that these must have come from the plants actually living in the peat swamps. To discover the parents, spores or pollen grains were usually isolated from the reproductive structures preserved as compression fossils in the shales between the coal seams. For example, the lepidodendrid cone *Lepidostrobus russelianus* is heterosporous, with two types of sporangia. One contains a few megaspores which, if found in sediments are called *Lagenoisporites rugosus,* while the other contains far greater numbers of microspores called *Lycospora.* From these and other similar studies it has thus been concluded that the coals dominated by *Lycospora* were derived from peats formed from the arborescent *Lepidodendron.*

The identity of the plant that produced *Densosporites* is far more problematical. This spore has been extracted from small, isolated cones called *Sporangiostrobus,* and recent work in Spain links these to small trees with a terminal tuft of leaves, although there is also a possibility that the parent plants were herbaceous lycopods forming a heath vegetation.

Evidence that the arborescent lycopods of *Lepidodendron* type were the pioneer colonisers of the swamp environments comes from a study of the seat-earths immediately below coal seems. These are dominated by *Stigmaria,* the underground rooting axes of the lepidodendrids, and many are found in life position. During periods of flooding, most of the swamp plants would have died and the decaying vegetation and underlying peat covered by the sands and muds carried in by the flood water. When the water level dropped, the sandy and muddy surfaces would have been invaded yet again by *Lepidodendron* and its allies. These in turn would have died and their remains accumulated as peat.

Direct evidence confirming the dominance of arborescent lycopods in the swamp forests comes from a study of peats that were petrified before they were compressed to form coal. Unfortunately, these calcium carbonate concretions, called coal balls, are not found in the Welsh coalfields but occur at one horizon in the north of England and throughout the North American Coal Measures. Studies of the American material suggest that for most of the late Carboniferous the swamps were dominated by *Lepidodendron*, with a lower level of pteridosperms and ferns, but as conditions became drier towards the end of the Period, tree ferns replaced the

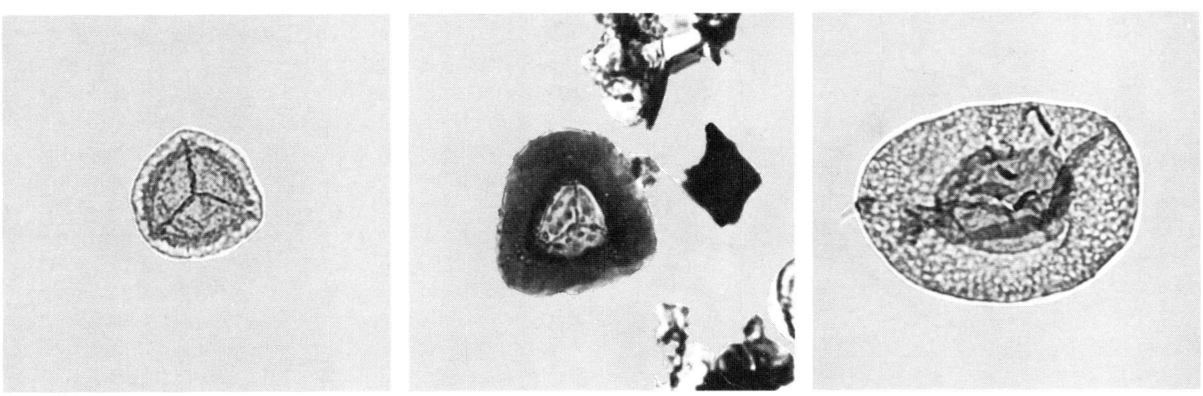

Spores from Upper Carboniferous (Coal Measures) plants. **(left)** Lycospora *sp.; this is the microspore of certain species of* Lepidodendron. **(centre)** Densosporites *sp.; this spore occurs in great numbers in certain coal seams and is probably of lycopod origin.* **(right)** Florinites *sp.; this type of pollen probably belonged to cordaites. (All x 500 approx.)*

lycopods as the most important forest plants.

A detailed analysis of sediments from between the coal seams provides evidence for plant communities that existed outside the forest swamps. Recent research on British coalfields suggests that a wide variety of plant communities dominated by pteridosperms and ferns grew on the river flood plains, while the margins of deltas and lakes were colonised by dense thickets of *Calamites*. The communities identified in these recent studies quite closely match the assemblages described by David Davies more than fifty years ago.

All these studies have concentrated on coastal vegetation. What about inland or upland vegetation? Uplands areas were places where erosion rather deposition occurred, and thus plants would rarely have became fossils under such conditions. Cordaites and conifers have traditionally been accepted as upland plants because they display certain anatomical characteristics, e.g. a leathery cuticle, which are often, although not exclusively, found in plants of dry environments. There is some evidence for this in the miospore assemblages from marine mudstone between certain coal seams which are dominated by pollen grains of cordaite type. It has therefore been postulated that the sea water flooded and killed the coastal vegetation so that the only remaining spore rain must have come from upland or inland areas. More recently, inland floras themselves have been found in the United States; they contain pteridosperms, cordaites and conifers.

Ecological interpretations that rely largely on the reconstruction of vegetational patterns obviously need to be based on the collection of large numbers of specimens, as shown by David Davies. Despite the destruction of numerous specimens today by mechanical mining methods, it still remains possible to make such large collections. For example, over the past few years some of the most beautifully preserved Coal Measures plants known from Wales have been collected from an open-cast site at Rhigos in Mid Glamorgan; the most important material from the site is in an extensive collection made by Mr. and Mrs. C. Taylor of Barry and subsequently presented to the British Museum (Natural History), where the associations are being studied in detail. Following the example of David Davies, the Taylors have demonstrated that amateur palaeontologists can continue to make significant contributions to palaeobotany in Wales.

Mesozoic and Tertiary floras

Major uplift and mountain building finally destroyed the Coal Measures environments about 280 million years ago and initiated a long episode of erosion throughout the Welsh area. During this time Permian rocks accumulated elsewhere in the northern hemisphere, with fossil floras indicative of rapid evolutionary changes associated with increasing aridity. By mid Permian times the vegetation of the much dryer lowlands was dominated by conifers. By the end of the Permian, the gigantic arborescent lycopods and horsetails characteristic of the Coal Measures swamps had disappeared, although herbaceous forms persisted. Of the gymnosperms, in addition to the conifers the pteridosperms survived into the Mesozoic although their leaves became much smaller and probably more leathery.

Mesozoic floras thus differ markedly from those of the Palaeozoic, but they are rarely recorded from Wales where only limited areas of Triassic and early Jurassic rocks are preserved as representatives of the sediments which accumulated following the Permian erosion. Fossils are rare in most of the local Triassic rocks, which were deposited in large inland salt lakes, probably surrounded by barren deserts. Some poorly developed stromatolites occur in the late Triassic Cotham Beds (Rhaetian, approximately 195-200 million years old) near Newport, Gwent, but at the same horizon on the opposite side of the Severn estuary in the Bristol district these structures are very well preserved in rocks known as the Cotham Marble. The algal laminated units are flat to gently domed or arranged as dendritic, arborescent drapes. Distinctive 'axial structures' in the laminates probably represent the sites of clumping of vertically oriented algal tufts, resulting from gliding algal filaments forming knots over which other filaments moved upwards towards the light; such clumping structures are commonly developed in modern oscillatoriacean-dominated mats.

Very slightly younger Rhaetian beds in South Wales, Gloucestershire and Somerset comprise brackish, lagoonal to marginal marine sediments that were deposited when the salt lakes were invaded by the sea. Fragmentary plants have been reported from the Welsh localities, but they are again preserved in better detail in the neighbouring areas; the most common but nevertheless remarkable genus is *Naiadita*, a small aquatic liverwort with elongated spear-shaped leaves borne on short, straight stems which are rarely branched.

Limited plant remains of unusual interest occur in Rhaetian or early Jurassic rocks deposited in fissures in the Carboniferous Limestone near Bridgend in South Wales. Cones, bark, wood and shoots of the conifer *Hirmerella* are the best preserved specimens, but fragments of other types also occur, mainly as the leaf genus *Pterophyllum*. The interest of the fossils lies in the fact that the majority of the specimens are preserved as charcoal in an uncompressed state; Professor T.M. Harris of

Polished surface of a piece of Cotham Marble. Upper Triassic (Rhaetian), Wickwar, near Chipping Sodbury, Gloucestershire (x 0.75 approx.)

Reading University has demonstrated convincingly that the plants were scorched and burned *before* becoming fossilised, providing evidence of a forest fire in early Mesozoic times, some 195 million years ago. Since the plant fragments were not completely reduced to ash it is likely that the fire passed quickly through the *Hirmerella* trees and scrub, resulting in the burning of twigs, leaves and plant debris, but only in scorching the more substantial timber.

Further important Rhaetian plants were discovered recently in a clay lens in excavations for the M4 motorway near Pyle in West Glamorgan. Together with some woody twigs, probably of conifer origin, and some fragments of an *Equisetum*-like plant, numerous fern leaves were found embedded in the clay. The discovery of the fern in such great abundance is particularly exciting as its only previous record in Britain is of a single waterborne fragment from the Rhaetian of Bristol. It has been identified by Professor Harris as *Clathropteris meniscoides,* a member of the extant tropical family Dipteridaceae. Professor Harris has described the plant as 'one of the noblest ferns of the Rhaetic and Lower Lias of the northern hemisphere.' A complete leaf must have looked like a giant parasol, 2 metres wide, which was split to the centre on one side, with the rest divided into segments, the divisions extending up to 20 centimetres from the stalk. Each segment was up to 14 centimetres wide. Intact leaves of such an enormous size have never been found. Professor Harris has suggested that the plants formed a riverside stand not far from where they were collected.

The plant record in the solid rocks of mainland Wales ends approximately 190-195 million years ago in the Lower Liassic (Jurassic) deposits of South Glamorgan and Gwent, from where a few fragmentary remains have been collected. Conifers are again present, but the best preserved specimens are the fronds of *Otozamites,* the foliage of a gymnosperm belonging to the Bennettitales. The fronds of *Otozamites* have two rows of elongated, linear leaflets with expanded bases attached to the upper side of a central axis; the veins of the leaflets are characteristically sub-parallel, radiating slightly from the base. The Bennettitales were an important group during Jurassic times, before becoming extinct by the end of the Cretaceous.

Early Cretaceous times (approximately 140-130 million years ago) witnessed the origin of the

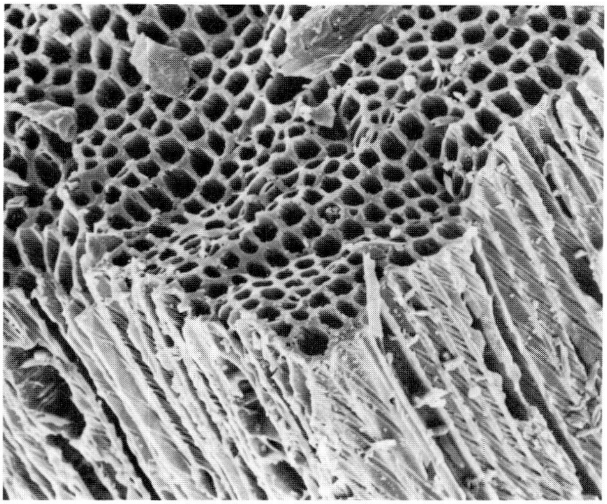

Scanning electron micrographs of early Mesozoic plants from fissures in the Carboniferous Limestone at Cnap Twt, near Bridgend, Mid Glamorgan. **(left)** *part of a leafy shoot of* Hirmerella muensteri? *(x 25 approx.);* **(right)** *a fragment of conifer wood showing detailed transverse and longitudinal structure (x 400 approx.)*

Clathropteris meniscoides; **(left)** *part of a frond (x 1 approx.);* **(above)** *details of the veins in the leaf (x 10). Upper Triassic (Rhaetian), near Pyle, West Glamorgan*

A *fragment of a conifer shoot with small, spirally arranged leaves. Jurassic (Lower Lias), Penarth, South Glamorgan (x 1.2)*

Part of the palm-like frond of Otozamites *cf.* obtusus. *Jurassic (Lower Lias), Penarth, South Glamorgan (x 1 approx.)*

flowering plants or angiosperms, but since no rocks of this age are preserved in Wales this important step in plant history is not recorded in the area. Succeeding Tertiary sediments are present on the mainland only as thin, local pockets of Palaeogene clays (about 50 million years old) in Dyfed and Gwynedd, but in the northern part of Cardigan Bay there are similarly aged sequences of terrestrial clays with sandy and silty partings that reach a thickness of at least 525 metres. In the core from the Mochras borehole off the coast of Harlech, plant debris is common throughout the Palaeogene sequence, and there are carbonaceous and lignitic partings occasionally thickening to a lignite bed. Together with a diverse assemblage of pollen and spores, the plants include mosses, conifer shoots, leaf cuticles and impressions, fruits, and seeds, and at some levels there are indications of rooting on old soils. As yet these plants from Mochras have not been studied in detail, but they will surely add further interest to the development of Welsh floras since Tertiary fossils from elsewhere provide evidence of the first appearance and establishment of many extant genera.

Quaternary floras

Deposits of glacial drift and outwash material from various parts of Wales help to provide evidence of the glaciations and interglacial periods of the 'Ice Age', approximately 1.6 million to 10 thousand years ago. The Quaternary record in Wales is dominated by evidence related to the youngest episodes of glaciation and ice retreat, but by incorporating data from neighbouring areas such as the English Midlands throughout the period a broad picture of climatic changes and associated floras can be drawn. Pollen, spores, mosses, peat and wood give a good indication of the advances and retreats of boreal and temperate floras in successive cold and relatively warmer periods during Pleistocene times. The boreal floras were characterised by trees such as *Pinus* (pine) and *Betula* (birch) in thin forests, with light-demanding herbs and shrubs as thin ground cover. Temperate deciduous woodlands established during the interglacials contained mainly *Quercus* (oak), *Corylus* (hazel), *Alnus* (alder), *Salix* (willow) and *Ulmus* (elm).

Following the 'Ice Age', the vegetational succession beginning in early Flandrian times (10 thousand years ago) records the development of temperate deciduous forests across Wales and their subsequent clearance by man. In addition, peat bogs were established at various periods up to about 1500 years ago, the evidence for these being particularly well exposed in deep excavations for docks and industrial sites along the South Wales coast over about the past hundred years. The submerged forests and peat beds that are familiar along many Welsh beaches at low tide provide a vivid picture of Flandrian sea level changes that destroyed the vegetation, although similar plants continue to live elsewhere in Wales today.

Acknowledgements

We are grateful to Mr. A. Harry and Mr. G. Watts for taking many of the photographs used in this article. In particular we thank Mr. F. Greenaway of the Photographic Unit, British Museum (Natural History) who photographed the specimens from the Taylor collection. For providing additional photographs and/or information we thank Dr. K.C.Allen, Dr. D.E.B. Bates, Dr. A. Beckett, Ms. R. Bennett, Dr. M. Butterworth, Dr. C.J. Cleal, Dr. C.J. Peat, Dr. J.B. Richardson, Mr. C. Shute, Prof. A.G. Smith, Mr. A. Stradling and members of the Porthcawl Museum and Historical Society, Dr. B.A. Thomas, Dr. R.H. Wagner and Dr. V.P. Wright. Prof. W.G. Chaloner kindly read the manuscript and offered suggestions for its improvement.

GEOLOGICAL TIME SCALE

ERA		PERIOD	AGE (Millions of years ago)	Summary of geological history of Wales
CENOZOIC	QUATERNARY	Recent or Holocene	0.01 – present	Coastal areas drowned by rising sea level caused by melting ice at the end of the Ice Age. Deposits of alluvium and peat, with further development of present drainage patterns and modern flora and fauna.
		Pleistocene	1.6 – 0.01	The 'Ice Age', with repeated glaciations and milder interglacial periods. The most recent major glaciation reached a maximum 18,000 years ago, and the last local ice left North Wales by 14,500 years ago. Modification of land forms by ice scouring and deposition of glacial drift. First evidence of Man in Wales from Pontnewydd cave, North Wales, about 200,000 years old.
	TERTIARY	Neogene	26 – 1.6	Prolonged, pulsatory uplift and erosion. Late Palaeogene and early Neogene terrestrial sediments known only from Mochras and locally in Gwynedd and south-west Dyfed. Early Palaeogene intrusive igneous rocks in north-west Wales. Basic landforms and drainage patterns established.
		Palaeogene	65 – 26	
MESOZOIC		Cretaceous	140 – 65	No rocks of this age known in Wales. Different opinions suggest either a persistence of terrestrial conditions, or that the Chalk sea covered much of the country.
		Jurassic	195 – 140	Lower Jurassic marine rocks known only from south-east Glamorgan and Mochras borehole, Harlech, with thick, younger sediments in Bristol Channel. Warm, shallow sea may have transgressed over Wales through the period but direct evidence is lacking.
		Triassic	230 – 195	Arid and semi-arid terrestrial conditions, with evidence of periodic flash floods. Dinosaur footprints and early mammal remains known from Glamorgan. Marine transgression in south-east Wales at the very end of the period.
PALAEOZOIC	UPPER	Permian	280 – 230	Uplift and mountain building (Hercynian Orogeny). Erosion across most of Wales.
		Carboniferous	345 – 280	Marine transgression early in period, leading to spread of warm, subtropical seas with rich coral/brachiopod faunas; extensive deposition of carbonate sediments that now form the Carboniferous Limestone. Regression in middle of period, with widespread deltaic deposits. Rich vegetation on coastal plains and deltas in late Carboniferous times; peat accumulated to form coal seams of the Coal Measures.
		Devonian	395 – 345	Uplift and mountain building (Caledonian Orogeny) continued from Silurian times, resulting in the deposition of the terrestrial Old Red Sandstone across most of Wales. Rapid diversification of land floras and non-marine fish faunas.
	LOWER	Silurian	435 – 395	Marine muds, silts and sands, with local carbonate sediments. Land with deltas across South Wales. Shallowing and retreat of the sea late in the period, with widespread onset of terrestrial conditions. Earliest fishes in Wales, and first vascular land plants appeared.
		Ordovician	505 – 435	Marine muds, sands, grits and local carbonate sediments. Evidence of extensive volcanicity in north, mid and south-west Wales. Fossil faunas increasingly diverse, including first corals.
		Cambrian	570 – 505	Transgression of sea into Wales, with deposition of grits, sands and muds. First abundant fossils, including earliest trilobites and brachiopods.
PRECAMBRIAN			4600 – 570	Oldest rocks in Wales dated at about 600-700 million years old, but are possibly considerably older. Evidence of intermittent marine conditions and volcanicity, with periods of folding, uplift and erosion. Earliest fossils from Wales are 'jellyfish' from Carmarthen of late Precambrian age.